HINDU ASTRONOMY

HINDU ASTRONOMY

W BRENNAND

GYAN

Published by

GYAN

Gyan Publishing House
5, Ansari Road
Daryaganj, New Delhi-110002
Phone: 011-47034999, 9811692060
E-mail: books@gyanbooks.com

Distribution Network
gyanbooks.com
India, USA, Canada, UK, Australia

ISBN: 978-93-6780-640-1 (PB)

First Published, 1896

2nd Impression 2024

Printed at: Gyan Press, Delhi.

Hindu Astronomy
Author: W BRENNAND

HINDU ASTRONOMY

BY

W BRENNAND.

WITH THIRTEEN ILLUSTRATIONS AND NUMEROUS DIAGRAMS

PUBLISHED BY CHAS. STRAKER & SONS, LTD,
BISHOPSGATE AVENUE, E.C

1896.

PREFACE.

It is perhaps expected that some reason should be given for the publication of this work, though it may appear inadequate. Force of circumstances, rather than deliberate choice on my part, impelled it; and, now that it has been accomplished, I cannot but feel how imperfect the production is. A lengthened residence in India led me to become interested in the study of the ancient mathematical works of the Hindus. This study was frequently interrupted by official duties, and much information acquired in its course has been for a time forgotten. Recent circumstances, and chiefly the interest displayed by my former pupils in a paper presented to the Royal Society on the same subject, has induced me to make an effort to regain the lost ground, and to gather together materials for a more extended work. Moreover, a conviction formed many years ago that the Hindus have not received the credit due to their literature and mathematical science from Europeans, and which has been strengthened by a renewal of my study of those materials, has led me also to a desire to put before the public their system of astronomy in as simple a manner as possible, with the object of enabling those interested in the matter to form their own judgment upon it, and, possibly, to extend further investigations in the subject. I have found far greater difficulties than I had anticipated from the fact that, although, no doubt, many Hindu writings exist which, if translated and consulted, would throw greater light upon the matter, yet comparatively few have undergone European investigation. I have been greatly assisted in my endeavours by the following books, from some of which I have made copious extracts, in order to present the views of others than myself.

The works of Sir W. Jones, Bailly's "Astronomie Indienne," and Playfaire's paper on it, in the "Transactions of the Royal Society of Edinburgh," Davis's "Essays in Asiatic Researches," Colebrooke's "Essays and Translations from the Sanscrit," Bentley's "Hindu Astronomy," references to Captain Wilford, Professor Max Muller, Ferguson's "Architecture," and other works; the Institutes of Akber, and the translation of the Siddhanta Siromani of Launcelot Wilkinson, C.S.

I have not entered at greater length into the mathematical knowledge of the Hindus than will be sufficient to show its general character, and that it was adequate for their requirements in the ordinary business of their lives, and for the purposes of their astronomy.

In the description of the Surya Siddhanta I have been indebted to the translation of this work from the Sanscrit, by Pundit Bapu Deva Sastri, of the Sanscrit College of Benares.

I take this opportunity of offering my thanks to my former pupils, who, after so many (23) years, still retain their attachment to me in my retirement, and especially to Rajah Rajendro Narayan Roy Chowdry of Bhowal, who have all taken the greatest interest in the progress of this work.

W. BRENNAND.

THE FORT,
MILVERTON, SOMERSET,
25th *March*, 1896.

CONTENTS.

PART I.

CHAPTER I.

PREHISTORIC ASTRONOMY OF THE ARYAN MIGRATING TRIBES.

CHAPTER II.

EARLY HINDU PERIODS.

CHAPTER III.

THE HINDU ECLIPTIC.

CHAPTER IV.

THE HINDU MONTHS, etc.

CHAPTER V.

THE RISHIS.

CHAPTER VI.

THEORY REGARDING THE CAUSE OF THE PLANETARY MOTIONS.

CHAPTER VII.

ARITHMETIC, ALGEBRA, AND GEOMETRY OF THE HINDUS.

CHAPTER VIII.

HINDU ASTRONOMICAL INSTRUMENTS.

CHAPTER IX.

EARLY HINDU ASTRONOMERS.

CHAPTER X.

RISE OF THE BUDDHIST HERESY, AND ITS EFFECTS ON HINDU ASTRONOMY.

CHAPTER XI.

PERIOD FROM THE RESTORATION OF THE POWER OF THE BRAHMINS TO BRAHMEGUPTA.

CHAPTER XII.

ASTRONOMICAL WORKS OF THE HINDUS.

PART II.

DESCRIPTION OF THE SURYA SIDDHANTA, WITH REMARKS AND EXPLANATIONS OF THE RULES.

Contents.

LIST OF ILLUSTRATIONS.

HINDU ASTRONOMY.

PART I.

CHAPTER I.

PREHISTORIC ASTRONOMY OF ARYAN MIGRATORY TRIBES.

In a zone of the Asiatic Continent, between 30 degrees and 45 degrees North, and from 30 degrees to 120 degrees East, about 900 miles in breadth, and nearly 4,000 miles in length (between Asia Minor and Africa on one side and the Pacific Ocean on the other), are some of the most extensive countries of the earth, and most productive and fruitful. Intermixed with them are many mountains, and deserts, high lands, and arid plains, with inland seas, lakes, and rivers. In some of these countries the people live in settled homes, engaged in agricultural pursuits; in others the Nomadic tribes, dwelling in tents, wander from place to place with their flocks, ever seeking fresh fields of pasture.

Such countries have, in Historic times, been the theatre of some of the most tragical events recorded in history, in which great nations have been the actors, and in which the empires of the Assyrians, the Medes, and the Persians, have each in turn risen, flourished, and long since been destroyed.

In times of peace the people have lived industrious lives, growing in wealth and numbers, acquiring the habits of civilization, cultivating the arts and sciences, and then have been swept away by some new wave of invading peoples, who have likewise given place to others.

It is a very reasonable presumption that, in Prehistoric times, similar eventful scenes have been enacted, and that, during the convulsions which overwhelmed the ancient great centres of civilization,

B

priests and rulers have gone forth from the intelligent classes, carrying with them their superior skill and learning, and have joined Nomadic tribes, becoming their leaders, and seeking other homes. There is evidence of several such great migrations from the districts above referred to, the discussion of which it is impossible to enter upon here. It will be sufficient to notice a few only of the opinions which writers on the subject have arrived at.

Sir W. Jones, in a series of discourses before the Asiatic Society of Calcutta in 1792, after a general survey of Asiatic nations, arrived at the conclusion that the Persians, the Indians, the Romans, the Greeks, the Goths, and the old Egyptians, all originally spoke the same language and professed the same popular faith, and this he conceived to be capable of incontestible proof.

Since that time, the theory thus propounded has gained strength. Many Oriental scholars have been engaged in a comparison of the languages, religions, customs, occupations, and the Mythologies of different nations of the earth, and their investigation has led them to the knowledge of a great body of facts, the explanation of which has become of great importance in the right interpretation of history. To all who have been thus engaged in the enquiry, it has brought the conviction that the Sanscrit, the Zend, and all European languages, are related to each other, and that the differences observed between them have arisen from the admixture of races, caused by great migrations from Central Asia.

Sir W. Jones considered it probable that the settlers in China and Japan had also a common origin with the Hindus and Persians, and he remarks that, however they may at present be dispersed and intermixed, they must have migrated from a central country, to find which was the problem proposed for solution. He suggests Irania as the central country, but he contends for the approximate locality rather than for its name.

Dr. O. Shrader, in his "Prehistoric Antiquities of the Aryan Peoples," has described the very various opinions, expressed by learned men of recent times, regarding the origin and homes of the Aryan races. Among the theories propounded, he mentions the opinion of Rhode, who endeavoured to discover the geographical starting point of the Zend people, in whom he comprehends Bactrians, Medes, and Persians. He observes traces of the gradual expansion of the Zend people, considering their starting point to be Airyana Vaejanh, followed by Sugdha, the Greek Σογδιανή (Suguda, Modern Samarkand). He further notes that "Eeriene Veedjo is to be looked for nowhere else than on the mountains of Asia, whence, as far as history goes back, peoples have perpetually migrated."

Dr. O. Shrader instances, as a proof of the close connection between the Indians and the Iranians, that they alike call themselves Arya, Ariya, and that, beyond doubt, India was populated by Sanscrit people from the North-west.

"There are clear indications," he further says, "in the history of the Iranian peoples, that the most ancient period of Iranian occupation was over before the conquest of the Medio-Persian territory, lying to the East of the great desert. From the nature of the case, it is just this Eastern portion of Iran, the ancient provinces of Sogdiana, Bactria, and the region of the Paropamisus, to which we must look in the first instance for the home of the Indo-Iranians."

Again, it appears to have been the opinion of Professor Max Muller, that "No other language (than the Hindu) has carried off so large a share of the common Aryan heirloom, whether roots, grammar, words, myths, or legends; and it is natural to suppose that, though the eldest brother, the Hindu was the last to leave the Central Aryan home."

Further, as Warren Hastings has remarked, when he was Governor-General of India, there are immemorial traditions prevalent among the Hindus that they originally came from a region situated in 40 degrees of North latitude.

The course taken by the great migration into India is supposed to be that which followed the ancient trade-route, and path of the

B 2

nations through Cabul, to the North-west of the Indus, and the progress is thus remarked upon by Dr. Shrader: "It is beyond doubt that India was populated by Sanscrit people from the North-west, a movement which is depicted in the Hymns of the Rig Veda as being in the course of progress. The Indians of this age, whose principal abode is to be looked for on the banks of the Sindhu (Indus), have as yet no direct knowledge of the Ganga (Ganges), which is only once mentioned in the Rig Veda*. Nor do their settlements seem to have reached as far as the mouth of the Indus, or as far as the Arabian Sea, at that time. The grand advance of the Indian tribes, Southwards and Eastwards, is mirrored very vividly in the different divisions and names of the seasons of the year in the more recent periods of the life of the Sanscrit language."

So, also, Bryant, in his "Ancient Mythology" (Vol. IV., 285), thus describes the effects of the emigrations referred to:—

† Upon the banks of the great River Ind, the Southern
Scutha dwell; which river pays
Its watery tribute to that mighty sea,
Styled Erythrean. Far removed its source,
Amid the stormy cliffs of Caucasus,
Descending hence through many a winding vale,
It separates vast nations. To the West
The Oritæ live and Aribes; and then
The Aracotii famed for linen geer.

---

* The most ancient writings of the Hindus are the Vedas, which are supposed by their followers, to be of divine origin. They are divided into four parts, each of which is a separate work, consisting principally of Hymns and Prayers, and, according to Colbrooke, they exhibit no trace of what must be considered the modern sects of Siva and Krishna.

They are named in the following order of their supposed antiquity:

The Rich or Rig-Veda written in Sanscrit Verse.

The Yajni-Veda in prose.

The Sama Veda on Chaunting, and the Atharva-Veda which consists of prayers and is believed to have had a later origin than the other three.

† Translated from the Greek of the Poet Dionysius.

Next the Satraidae; and those who dwell
Beneath the shade of Mount Parpanisus,
Styled Arieni. No kind glebe they own,
But a waste, sandy soil replete with thorn.
Yet are they rich; yet doth the land supply
Wealth without measure. . . . . .

. . . .

. . . . To the East a lovely country wide extends—
India; whose borders the wide ocean bounds.
On this the sun new rising from the main
Smiles pleased, and sheds his early orient beam.
The inhabitants are swart; and in their looks
Betray the tints of the dark hyacinth,
With moisture still abounding; hence their heads
Are ever furnished with the sleekest hair.
Various their functions: some the rock explore,
And from the mine extract the latent gold.
Some labour at the woof with cunning skill,
And manufacture linen; others shape
And polish ivory with the nicest care.

. . . .

. . . .

Nor is this region by one people held;
Various the nations, under different names,
That rove the banks of Ganges and of Ind.
Lo! where the streams of Acasine pour,
And in their course the stubborn rocks pervade,
To join the Hydaspes! Here the Dardans dwell,
Above whose seat the River Cophes rolls.
The sons of Saba here retired of old;
And hard by them the Toxili appear,
Joined to the Scodri. Next a savage cast—
Yclep'd Peucanian. . . . . .

. . . .

To enumerate all who rove this wide domain
Surpasses human power. The Gods can tell—
The Gods alone, for nothing's hid from Heaven.
Let it suffice if I their worth declare:
These were the first great founders in the world—
Founders of cities and of mighty states,
Who show'd a path through seas, before unknown,
And when doubt reigned and dark uncertainty,
Who rendered life more certain. They first viewed
The starry lights, and formed them into schemes.
In the first ages, when the sons of men
Knew not which way to turn them, they assigned
To each his just department; they bestowed
Of land a portion and of sea a lot,
And sent each wandering tribe far off to share
A different soil and climate. Hence arose
The great diversity, so plainly seen
'Mid nations widely severed. . . . .

But it was not only Southward to India that the Nomads of Central Asia migrated. They spread Westward and Eastward.

Du Halde, in his account of the Jesuit Missions in China, given in his description of the Empire, says:—"It is a common opinion of those who have endeavoured to trace the origin of the Empire, that the posterity of the sons of Noah, spreading themselves over the Eastern parts of Asia, arrived in China about 200 years after the Deluge, and settled themselves in Shen-Si." He supposes the Flood to have happened in the year 3258 B.C., preferring the account of the Septuagint to that of the Vulgate. He then rejects the dates of the first Emperors of China, which are given in "Annals of the Chinese Monarchs," as being uncertain, and as involved in some degree of obscurity, and estimates that Yu the Great—the first Emperor of the Dynasty called Hya—began to reign in the year 2207 B.C. As an eclipse of the sun happened in the year 2155 B.C., which has been astronomically verified, and which is also recorded in the Chinese history, it is considered to be demonstrated that China must have been peopled long before that time, and that the date of the first Emperor was about the year 2327 B.C. There is a great diversity of opinion regarding the introduction and use of the first Chinese cycle of 60 years (which they brought with them from the West, and which was also a common cycle in India and Chaldea), some placing it at 2757 B.C.; and this epoch is that which is generally accepted in China at the present day.

Now, at the epoch mentioned (2757 B.C.), when the tribes migrating Eastward from some central country in the West, were on their way to their new homes in China, it is an astronomical fact that α Draconis was their polaris. It is a star of the third magnitude, and would be seen by them, apparently a luminous point, fixed at the North Pole, between 30 and 40 degrees in altitude. They would also have seen other stars of the Constellation Draco, describing small circles about this point, greater and greater, according to their distances from it. Night after night, as they tended their flocks, the same phenomena would be witnessed by them, and must have been vividly fixed in the memory of all. Those who have any acquaintance with Chinese history know how much the dragon is held in veneration by them, it being the symbol of royalty, emblazoned on their temples, their houses, and their clothing.

In the year 2800 B.C., α Draconis was only 10 minutes from the Equinoctial Pole, and being then in the Solstitial Colure, it must have impressed upon observers of that period, in a higher degree, the sacred character in which the times of the solstices and the equinoxes were held in their former homes. The Chinese religion then resembled that of the Indian Vedas, showing an affinity between the two races. On four mountains at the extremities of the Celestial Empire, four altars were placed, on which offerings and oblations were laid, with prayers, and were the homage paid to t-hyen, or the sky, which was considered an emblem of the Supreme Being, the Creator and Ruler of the Universe. Four solemn sacrifices were at that time ordered to be offered on the Eastern, the Southern, the Western, and the Northern mountains, at the equinoxes and solstices in regular succession. Similarly, in India, the Brahmins are enjoined, in the Institutes of Menu*, to make sacrifices in honour of the Lunar mansions, and holy rites were observed every three months, at the equinoxes and at the winter and summer solstices.

In this worship of the Most High we see some resemblance also to the prayers and sacrifices of the Nomadic Abraham, who was himself a wanderer from the house of his fathers in Ur of Chaldea, as described in the Hebrew Scriptures.

The emigrating tribes, who thus, undoubtedly, in Prehistoric times, went forth Westward, Southward and Eastward from their Central Asiatic homes, carried with them evidences of their common origin. For example, they had the same religious belief in one Supreme Being, the Creator and Supreme Ruler of the Universe, to whom prayers and sacrifices were offered.

They had the same days of the week, over which the sun, the moon, and the five planets were supposed to have been appointed rulers, in successive order, in accordance with their respective names.

They had the same divisions of the Ecliptic, into twelve parts or signs of the Zodiac, corresponding with the twelve months of the year, the sun moving through the successive signs, during successive months.

There was also, among the more intellectual classes of all these wandering Asiatic races, another division of the Ecliptic, into 28 parts, forming the extent of the same number of constellations or

---

* "The Institutes of Menu" is a treatise on religious and civil duties prescribed by Menu, the son of Brahma, to the inhabitants of the earth. It is a work in Sanscrit, and, next to the Vedas, of the greatest antiquity. A translation into English is given in the works of Sir W. Jones, vol. VII. Reference will be made to it further on.

Asterisms, being the spaces through which, in succession, the moon travels daily, in its monthly course round the heavens. This system of Lunar constellations, though common to several Eastern countries, has different names in each. In the Indian astronomy the Asterisms are called Nacshatras; in the Chinese they are designated Sieu; in the more central parts of Asia they had the name of Manzils, *i.e.*, Lunar mansions or stations. They are, however, less known among Western nations. The Egyptians possessed them at a comparatively late date, but made little or no use of them, and it does not appear that they have had any place in Grecian astronomy. It was the diligent use which the Hindu astronomers made of these Nacshatras, in the progress of their astronomy, that gave them their superiority over all other ancient nations, and in so far as they permanently introduced this Lunar system of division in their Ecliptic into their astronomy, it would appear to be characteristically different from our modern system.

It has, however, been a question raising much discussion amongst the learned for many centuries, as to who were the original inventors of the Celestial sphere, which has descended to us from the Greeks, with their vast system of mythical fables.

The Solar Zodiac, indeed, with figures representing the 12 signs, has been in use in all historical periods, having nearly the same characteristics among the Greeks, the Egyptians, the Persians, the Hindus, the Chaldeans, and the Chinese.

It is reasonable, therefore, to suppose that the idea of the Celestial Sphere and of the Solar Zodiac was a common possession of all the migrating tribes referred to by Professor Max Muller, at times before they left their central homes.

The Nomadic tribes of Asia, who watched their flocks by night, must, as they themselves wandered over vast plains in search of fresh herbage, have had abundant opportunities of observing the Sidereal Sphere, which was apparently in incessant motion. Night after night, with an unobstructed view, the same stars would be seen to rise in the East, and to pursue an even course through the sky, but to set a little earlier each succeeding night. From childhood upwards, every individual of the tribes must have become familiar with the forms in which the stars were constantly presented to their view. What, then, would be more natural than that they should speculate regarding the nature of Celestial orbs; that in fancy they should have pictured to themselves outlines among them; and in imagination given them the forms of objects with which they were most familiar?

The ram, the bull, the goat and kids, the virgin reaper, or gleaner of corn, the archer, who, in defence of the flock, must have had conflicts with the lion; the bearer of water to the cattle; the crab and fish of the lakes and rivers, which they frequented, the poisonous scorpion, and the balance designating the time when the days and nights were equal; all indicate the common objects of the wild and restless people of the plains, and emphasise the probable fact that the signs of the Solar Zodiac originated with the Prehistoric Nomads of Central Asia.

Time, during the long night watches, could only be known by the motions of the luminous orbs of the sphere, and the bright stars, at their rising and setting, were such familiar objects, and so generally known, that they would be referred to with the same ease and confidence with which, in modern times, we refer to a watch or a clock. The rising of the sun began the solar day, and a bright star, rising at the same moment, is often referred to as marking a particular time of the year, in connection with some other event expected to happen at the same time. Thus the heliacal rising of Sirius was connected with the inundation of Egypt by the rising of the Nile, about the time when Sirius rose with the sun. The births of children were marked by some star rising at the same time with the sun, or connected with that point of the Ecliptic which was in the horizon at the same moment, thus constituting the child's horoscope, by which, in after times, it was believed the astrologer could foretell all the events that would happen in the child's life.

Moreover, it may be readily supposed that the fertile imagination of the Eastern story-teller would find, in the starry sphere, the means of giving a celestial locality to departed heroes and to other objects connected with the tales and traditions of his tribe; illustrations, which would give to his listeners a greater interest in his romances during the tedious hours of the night, whilst they were tending their flocks.

In this way it is not improbable that all the 48 ancient constellations received their names, and that the rising and setting of particular stars with the sun were noted and connected with events in their lives, which may have been the origin of many superstitions prevalent in later times, and, in some cases, may have been, amongst the Greeks, the origin of their poetical legends.

The approach of the rising sun would be indicated when the night watch was nearly over, by the fading and final disappearance of the smaller stars, in the increasing dawn. Long shadows would be thrown out by the tent poles towards the West, which would shorten and change their directions, as the sun ascended higher in the sky, and at his highest point they would be shortest, and at the moment directed to the North, indicating that the sun was in the meridian, and that it was noon. Then the changes would begin in a reverse order, the shadows lengthening, but turning still in the same direction towards the East, disappearing when greatest at sunset.

These circumstances no doubt originated, in its earliest form, the sun dial (a vertical style on a horizontal plane) the same form as described in the Surya Siddhanta of the Hindus. and in the description given by the Jesuit missionaries, as being in use in the Chinese observatories.

And again, to the Nomadic tribes in Prehistoric times may be ascribed the simple discovery that when the sun in summer rose at a point of the horizon nearest to the North, and furthest in amplitude from the East, its place in the Ecliptic was the Northern solstice, which then rose with the sun, and that it was then a time of most solemn import—a day when prayers and sacrifices should be offered to the Almighty. This, indeed, was a religious practice observed among all the emigrating tribes both at the solstices and the equinoxes.

That the bull should have been held to be a most sacred animal in Prehistoric times among the migrating tribes, and afterwards especially revered in Egypt and India, is a circumstance that would appear to have its explanation in the fact that, between the years 2426 B.C. and 266 B.C., the equinox was retrograding through the Constellation Taurus.*

So, also, about the time when the various tribes were migrating from their central abode, the bright star Cor Leonis must have been an interesting object to the primitive astronomers of that period, for this star was then at or near the summer solstice, and a parallel of declination through it in the year 2305 B.C. might be

---

* Among ancient entablatures which are carved in rocks we have observed figures with the head and horns of Bulls.

The Egyptians undoubtedly worshipped one of these animals at their City of Pharbethus.

When the Apis died it was put into a coffin and interred in the temple of Ser-Apis.

The mino-taur, the Taurus Lunaris of Crete, was represented as a man with the head of a bull:

The Bull's head was esteemed a princely hieroglyphic, and Astarte, it is said, placed the head of a bull upon her head as a royal emblem.

Mountains, places and peoples are named Taurus, Taurica, Taurini, Taurisci, Tauropolis, Tauropolium.

Tours, in Gaul, was called ταυροις. Many other instances may be collected from China, Japan and India.

In India the Brahmini Bull wanders freely through the towns and villages, a mendicant receiving doles of rice from village shopkeepers. It is often made away with by Mahomedan butchers, and such desecration has occasioned frequent encounters between the Hindus and Mussulmen.

properly termed then the "tropic of Leo." It may be owing to this fact that the Persian priests of Mithra, clothed in the skins of lions at the Mysteries called Leonticæ, were named lions.

Again, at a period of 120 years before this time, the solstice was at the point dividing the two constellations Leo and Virgo, which circumstance probably gave rise to the enigma of the Sphinx, the Egyptian image, which, in the greater number of cases, had the head and breast of a virgin and the body of a lion (implying a doubt whether the virgin or the lion was most to be adored).*

---

* Sphynxes are stupendous monuments of the skill of the Egyptians. The largest and most admired of these, like the pyramids, seems partly the work of nature, and partly that of Art, being cut out of the solid rock. The larger portion, however, of the entire fabric, is covered with the sands of the desert, which time has accumulated round these master-pieces of other days, so that the pyramids have lost much of their elevations.

The number of Sphynxes found in Egypt, besides their shape, seems to countenance the oldest and most commonly received opinion, that they refer to the rise and overflow of the Nile, which lasted during the passage of the Sun through the constellations Leo and Virgo: both these signs are, therefore, combined in the figure which has the head of a Virgin and the body of a Lion.

The largest Sphynx was imagined also, as Pliny affirms, though with what reason does not appear, to have been the sepulchre of King Amasis. It having been considered that time must have effected revolutions, in respect of the signs themselves, of which these structures were supposed to be symbols, as regards the rising of the river and the order of the months, it has been more recently concluded that the Sphynxes were mysterious symbols of a religious character not now to be unravelled.

According to Herodotus the periodic inundations commence about the end of June and continue till the end of September.—*Encyclopædia Metropolitana, IX.*, 209.

The Theban Sphinx has the head and bosom of a girl, the claws of a lion, the body of a dog, the tail of a dragon, and the wings of a bird.

Count Caylus thinks that the Sphinx was not known in Greece, but by the story of Œdipus, and then it appears in the same manner as when proposed in the enigma.

The Egyptian Sphinx, says Winckleman, has two sexes, the Andro-

The revolutions of the moon, and the phases it daily assumed, increasing till, at the full, it was a circle, and then decreasing till it finally disappeared at conjunction, could not fail to be objects of great interest to the wandering tribes. The synodic period, or the time between one new or full moon to the next, a little more than 29½ days, but reckoned by them as 30 days, must have been the means of marking the extent of each constellation in their Solar Zodiac. And 12 such periods of the new moon constituted, in the earliest times, the solar year of 360 days, when it was assumed that the sun had completed its course through the ecliptic circle, thus making the diurnal motion of the sun to correspond with one degree of arc.

The divisions of the Zodiacs of all the countries before mentioned were the same, each constellation extending over 30 degrees of arc, the irresistible inference being that they all derived their Zodiacal division from a common ancestry.

In the Zodiacs of different Western countries, the same Solar constellations were, in general, represented by the same figures of animals and other objects, that of the Persians differing from them only by representing the twins as two kids. Moreover, the abbreviations are nearly the same, though the figures differ from the general type, as will be seen by reference to the two accompanying plates. (Plates I. and II.)

The figures, however, of the same Solar constellations of the Chinese Zodiac, are, with the exception of two, entirely different.

Sphingis of Herodotus with the head of a female and male sexual parts. They are found with human hands, armed with crooked nails, with beards; the Persea plant upon the chin, horses tails and legs; veiled, the sistrum, &c.

Plutarch says that it was placed before the temples to show the sacredness of the mysteries.

In Stosch it holds in the mouth a mouse by the tail, has a serpent, before her a caduceus.—*Fosbroke I.*, 153.

PLATE I.

THE ANCIENT ZODIAC OF EGYPT

WITH THE ORIGINAL ASTERISMS.

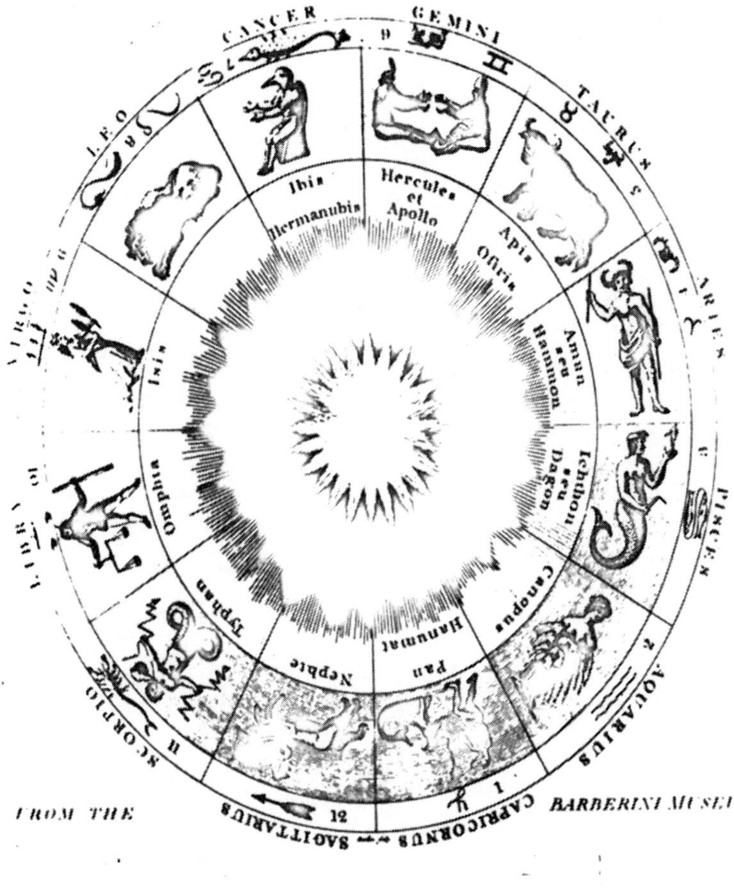

FROM THE BARBERINI MUSEUM

PLATE II.

ORIENTAL ZODIACK.

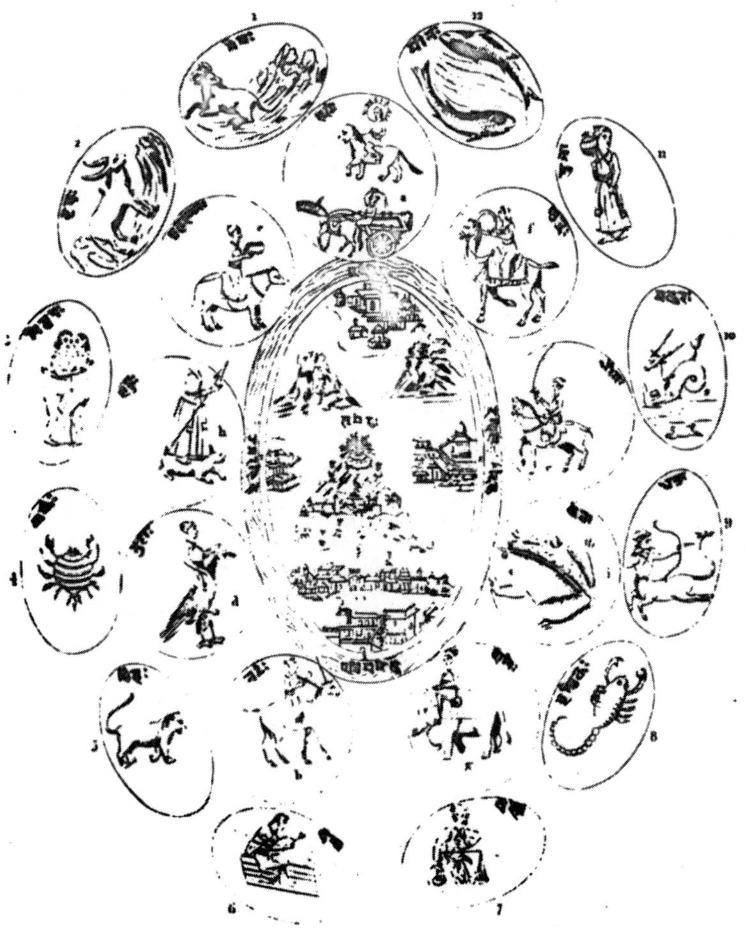

From 1 to 12 are the 12 Signs. a The Sun b The Moon c Mars d Mercury e Jupiter f Venus g Saturn h Dragons Head or Ascending Node. Dragons Tail or Descending Node. The center is the Earth surrounded by the Sea, marked with the four Cardinal Points E.W.N.S. W.X.Y.Z.

Only the 2nd and the 8th (the ox and the sheep) may be supposed to resemble two corresponding animals in the others. Out of the 10 remaining figures only five are met with among those which represent the 48 constellations of the ancient celestial sphere. These are the dragon, the serpent, the horse, the dog, and the hare.

The following table shows a comparison of the Chinese signs with those used by other nations:—

The constellations of the Chaldean Zodiac represented by—	*Hindu names.*	According to Kempfer constellations in the Chinese Zodiac are represented by—
1 The Ram	(Mesha)	1 The Mouse
2 „ Bull	(Vrisha)	2 „ Ox or Cow
3 „ Twins	(Mit'huna)	3 „ Tiger
4 „ Crab	(Carcati)	4 „ Hare
5 „ Lion	(Sinha)	5 „ Dragon
6 „ Virgin	(Canya)	6 „ Serpent
7 „ Balance	(Tula)	7 „ Horse
8 „ Scorpion	(Vrishchica)	8 „ Sheep
9 „ Archer	(Dhanus)	9 „ Archer
10 „ Goat	(Macara or sea monster)	10 „ Cock
11 „ Water-bearer	(Cumbha)	11 „ Dog
12 „ Fishes	(Mina)	12 „ Boar

It is, however, to be observed that, although there are these remarkable differences in the names and figures of the signs employed by the Chinese, yet the number of the whole is the same as that employed by the other nations. The differences, therefore, it may be assumed, were those naturally arising from the locality in which the tribes found themselves.

* At Achmin or Echmin, the ancient Chemnis or Panopolis of the Greeks near Ptolemais Hermii, is a town in which there is a temple of which the stones are scattered about, others transferred to a Mosque where they are placed without regularity or taste. On one of them may be traced four concentric circles in a square, the innermost of which contains the Sun; of the next two, one contains 12 birds, the other 12 animals almost effaced, which appear to be the signs of the Zodiac.

The fourth has 12 human figures, which M. Savary imagines to represent the 12 Gods, the 12 months of the year and the 12 signs of the Zodiac.

It is, at any rate, certain, notwithstanding the differences just alluded to, that all the migrating tribes carried with them a conception of the division of the sun's path into 12 equal parts, which formed the extent of each of the Solar constellations, and so the Solar Zodiac became one of the foundations on which all their astronomical systems were constructed.

But amongst the primitive astronomers, there were, apparently, at least two distinct sects in each tribe, the one adopting the Solar Zodiac, which had animals principally for its symbols, and another sect which assumed a division of the Ecliptic into 28 parts, corre-

---

The four seasons occupy the angles of the square, on the side of which is discernible a globe with wings.

It seems probable that this temple was dedicated to the Sun, and that the whole of these hieroglyphics mark his passage into the signs of the Zodiac and his annual revolutions, Longitude 31° 45, Latitude 26° 35.—*Encyclopædia Metropolitana.*

According to Macrobius the Signs of the Zodiac originated with the Egyptians, though the jealous Greeks laid claims to the invention.

The Ram was assimilated to Jupiter Ammon; the Bull to Apis; the Gemini to the inseparable brothers Horus and Harpocrates, who became Castor and Pollux; Cancer to Anubis, who was changed to Mercury by the Greeks and the Romans; Leo to Osiris, emblem of the sun; Virgo to Isis, converted into Ceres; Libra did not exist in the Egyptian Zodiac, and its place was occupied by the claws of the Scorpion; Scorpio was converted to Typhoon, and became the Greek Mars; Sagittary was made Hercules, the Conqueror of Giants (Macrobius 1, 20); Capricorn was Mendes, the Egyptian Pan; Aquarius, Cornopus; Pisces, Nephlis, the Greek Venus.—*Fosbroke's Encyclopædia of Antiquities, Vol. I., p.* 222.

The Ram was an animal consecrated to the Egyptian Neitha, a goddess who presided over the Upper Hemisphere, whence Aries was dedicated to her.

Cancer was the Crab who stung Hercules in the foot to prevent his killing the Hydra, and transformed by Juno after he had trodden it to death to the Zodiac.

Capricorn was either the Amelthian Goat or Pan, who metamorphosed himself, through fear of the Giant Typhon, into a goat in the upper part,

sponding with a like number of constellations, marking the daily progress of the moon through them, and which were designated in the ancient astronomy as the "Lunar Mansions." It would appear that the Solar Zodiac was made the principal foundation of the Western Astronomies of Egypt and Greece, and, in connection with its symbols, their respective systems of Mythology were formed; but in the more Eastern countries (especially in India), in the earliest ages, although the Solar Zodiac was retained, a preference would seem to have been given to the Lunar mansions, from which were constructed the Lunar and Luni-Solar years.

It is, moreover, probable that the titles given to the two ancient races of Indian princes, both of the posterity of Menu, and called "the Children of the Sun, and the Children of the Moon" (who reigned respectively in the cities of Ayodha or Audh, and Pratishthana or Vitra), had their origin in the astronomy of the two sects which severally adopted, for the foundation of their doctrines, the Solar Zodiac and the Lunar Asterisms respectively.

The tribes which wandered further eastward also carried with them to their final settlements in China the two methods of dividing the Ecliptic above described. The former of these, with animal symbols, in China, differed widely, as has been shown, from that adopted by Western astronomers, and the latter (the division of

---

and a fish in the lower, which so surprised Jupiter that he transported him into the sky.

Leo is the Nemean Lion.

Sagittary is according to some the Centaur Chiron, according to others Orocus, whom the Muses requested after death to be placed among the Signs.

Scorpio, that insect whom Jupiter thus honoured after its battle with Orion.

Pisces are the fish which carried on their backs Venus and Cupid, when they fled from Typhon.

The Bull, the oldest Sign is taken from the deep Oriental Mythology.

Aquarius is Ganymede thus elevated by Jupiter.—*Encyclopædia Metropolitana.*

C

28 Lunar Asterisms, stations, or mansions) bore a close resemblance to those of other Eastern nations, insomuch that Bentley, after examining the astronomy of the Chinese, says:—"I found the Chinese were not only far behind the Hindus in the knowledge of astronomy, but that they were indebted to them in modern times for the introduction of some improvements into that science, which they themselves acknowledge." Yet the Chinese Asterisms differed greatly in point of space occupied by each; for, as Bentley says:—"With respect to the Lunar mansions of the Chinese, they differ entirely from those of the Hindus, who invariably make theirs to contain 13° 20′ each on the Ecliptic; whereas the Chinese have theirs of various extents from upwards of 30 degrees to a few minutes, and markeo by a star at the beginning of each, which makes them totally differ from the Hindus." [See Plate III.]

The Arabs are supposed by him to have communicated their Asterisms to the Chinese. On comparing these two systems he found that "13 out of the whole number, which consists of 28, were precisely the same, and in the same order, without a break between them; consequently there must have been a connection between them at some time."

The question then arose whether the Chinese borrowed from the Arabians or the Arabians from the Chinese. Bentley says he mentioned the circumstance to a learned Mahomedan, in the hope of getting some information, and his reply was, "that neither the Chinese borrowed from the Arabs, nor the Arabs from the Chinese; but that they both had borrowed from one and the same source, which was from the people of a country to the North of Persia, and to the West or North-west of China, called Turkistan. He observed that before the time of Mahomed the Arabs had no astronomy, that they were then devoid of every kind of science; and what they possessed since on the subject of astronomy was from the Greeks. To which I replied that I understood the mansions of the moon were alluded to in the Koran, and as the Greeks had no Lunar mansions in their astronomy, they could not come from them. He said the mansions referred to in the Koran were uncertain, that no one knew what particular star or mansion was meant, and, therefore, no inference could be drawn that any of those now in use were alluded to. Here our conversation ended."

As the similarity between the Arabian and the Chinese Asterisms apparently gave rise to the surmise that the latter were borrowed from the former, it may be advantageous at this point to examine the Arabian system.

An account of the Lunar mansions, called Arabian, was given by Dr. Hyde, Librarian of the Bodleian Library, in a work entitled "Ulug-Begh Tabulæ Stellarum Fixum," translated from the Persian, Oxon, 1665; and from this work we have the names of the 28 Lunar Constellations.

Costard (in his "Chaldaic Astronomy," Oxon, 1748) was of opinion that the Lunar mansions of the Arabians were derived immediately from the Chaldeans.

The greater probability, however, as has been suggested, is that the Lunar mansions found to be a portion of the Chinese, the Indian, the Arabian, and the Chaldean astronomy, all had a common derivation from the emigrated peoples of Central Asia.

Ulug-Begh was a chief or monarch of the Tartars. He was devoted to the study of astronomy, and in his capital of Samarcand he had an observatory, with a quadrant 180 feet high, with which he made good observations. His principal work was a catalogue of the fixed stars, composed from his own observations in 1437 A.D., said to be so exact that they differed little from those of Tycho Brahe. The latitude of Samarcand is put at 39° 37′ 25″ N.

The names of the Lunar Asterisms, according to Ulug-Begh, are given in the following order by Dr. Hyde, with his observations, and remarks on them by Costard:—

c 2

Names and significations of the Manazil-Al-Kamar, or Mansions of the Moon.

1. Al-Sheratau: They are the two bright stars in the head of Aries.
2. Al-Botein: From betu, venter; they are small stars in the belly of the Ram.
3. Al-Thuraiya: From therwa, multus, copiosus, abundans; they are the Pleiades.
4. Al-Debasan: Properly the Hyades, but generally applied to the bright star in the head of Taurus, called, in Arabic, Ain-Al-Thaur, or the Bull's Eye.
5. Al-Hekah: The three stars in the head of Orion.
6. Al-Henah: Two stars between the feet of Gemini.
7. Al-Dira: Two bright stars in the heads of the two Gemini.
8. Al-Nethra: The Lion's Mouth.
9. Al-Terpha: The Lion's Eyes.
10. Al-Giebha: The Lion's Forehead, or, according to Alfragani, four bright stars in Leo, one of which is Cor-Leonis.
11. Al-Zubra: Two bright stars, following the Lion's Heart.
12. Al-Serpha: The Lion's Tail.
13. Al-Auwa: The five stars under Virgo.
14. Sinak-Al-Azal: The spike of Virgo.
15. Al-Gaphr: The three stars in the feet of Virgo.
16. Al-Zubana: The Balance.
17. Au-Iclil: The Northern Crown.
18. Al-Kalb: The Scorpion's Heart.
19. Al-Shaula: Two stars in the tail of Scorpio.
20. Al-Naaim: Eight bright stars, four of which lie in the Milky Way, and four of them out of it; those in the Milky Way are called Al-Warida, or Camels Going to Water; those out of it, Al-Sadira, or Camels Returning from Water.

Plate III.

COMPARISON OF THE LUNAR MANSIONS.

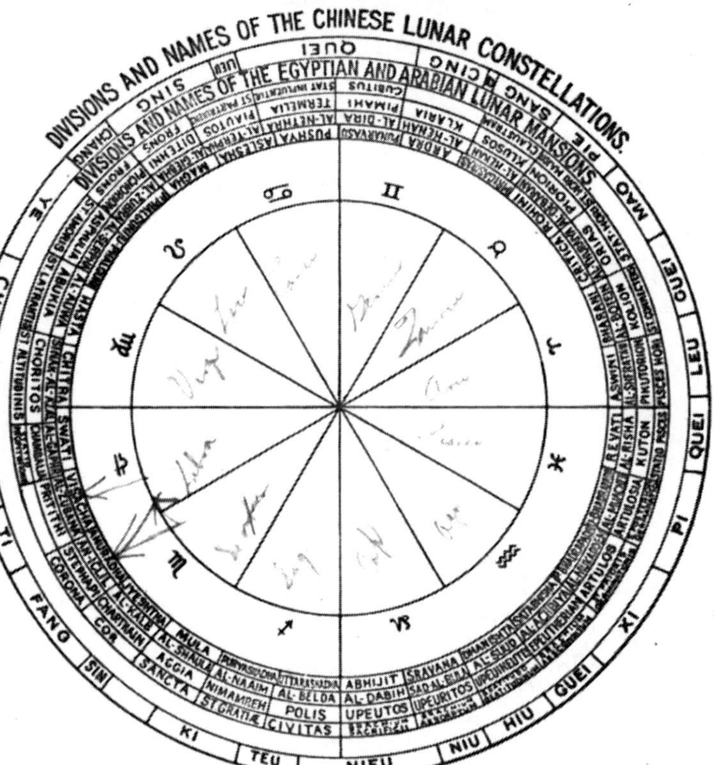

21. Al-Belda: "Quod urbem, oppidumve denotat," says Dr. Hyde: According to some Arabian astronomers, it means six stars in Sagittarius, where is the sun's place the shortest day of the whole year. According to others, it is a portion of the heavens entirely destitute of stars, succeeding the Al-Naaim. Why a vacant space should be called a town or city the learned commentator has not informed us. I should rather think the name alluded to the extraordinary number of six stars, which crowd this Lunar abode in Sagittarius.

22. Al-Dabih: Four stars in Capricorn.

23. Sad-Al-Bula: The sixth star in Ulug-Begh's table of Aquarius; it is probably that marked γ by Bayer.

24. Al-Sund: Two stars in Aquarius, marked β and ξ by Bayer.

25. Al-Achbiya: Four other stars in Aquarius, marked by Bayer γ. ζ. η and θ.

26. Al-Phergh—Al-Mukaddem: Two bright stars, of which the Northern one is called the Shoulder of Pegasus. They are marked by Bayer α and β in his table of that constellation.

27. Al-Phergh—Al-Muacher: Two bright stars at a distance from each other, following Al-Phergh, Al-Mukaddem. One is in the head of Andromeda, and the other is Bayer's γ. in the extremity of the wing of Pegasus.

28. Al-Risha: The fumes, the cord; that is of the fishes. In Alfragani this Lunar Mansion is denominated Batu-Al-Hut, venter piscis, and it is said to mean the stars of the Northern Fish.

In Plate III. a comparison is instituted between the 28 Lunar mansions of the three countries, China, Egypt, and Arabia, with the 28 Indian Nacshatras of $12\frac{6}{7}$ degrees each, into which the Indian Ecliptic was supposed to be equally divided.

The different systems have been plotted from tables reduced, and so as to make the beginning of the longitudes in each, the same point, *i.e.*, the beginning of Aswini.*

The longitudes of the Arabian Manazil-Al-Kamar—are taken from a table by Martinius, and compared with those of a table by Bentley, which he deduced from the places of stars in Dr. Morrison's Chinese Dictionary. The names of these Chinese Asterisms are those given by Du-Halde, and copied into the astronomy of Dr. Long.

The longitudes of the Arabian Manazil-Al-Kamar—are taken from a table calculated by Costard, who was at great labour in reducing the latitudes and longitudes from the observations of Ulug Begh. He observes that, whatever opinion may be entertained relative to their antiquity, they must, at all events, be older than the time of Mahomed, because the Lunar stations, as well as the Solar, are alluded to in the Koran in the following passage:—"Posuit Deus Solem in splendorem, et Lunam in turnen; et disposuit Eam in Statione ut sciretis numerum annorum."

The names and longitudes of the Egyptian system are as recorded by Bentley, who says he took them from the Lingua Egyptiaca of Kircher. Bentley remarks that the Egyptians make the Equinoctial Colure to cut the star Spica Virginis, and, in consequence, he considered the table to have an epoch of 284 A.D.

Another feature, in which the astronomy of Eastern nations appears to be connected, may be recognised in the cycle of 60 years adopted in each.

This cycle of 60 years was brought into India by some of the immigrant tribes, and was afterwards known as the Cycle of Vrihaspati, *i.e.*, of Jupiter. It is a combination of two cyles, a cycle of five years, from the Jyotish (or Astronomy) of the Vedas,

* A description of the Indian Ecliptic, subsequently adopted with 27 asterisms, is intended to be given hereafter, together with an explanation of the origin here alluded to as "the beginning of Aswini."

and the sidereal period of the planet Jupiter, which was at first reckoned to be 12 years, but was afterwards found by the Hindus to be 11. 860962 years. According to Laplace, the mean sidereal period is 11.862 Julian years, or .138 of a year short of 12 years, an error of about 8¼ months in 60 years, and would, therefore, require periodic correction.

It has already been mentioned that Chinese History and the Annals of the Chinese Emperors were written by reference to cycles of 60 years. Such a period of time, moreover, was in common use in Chaldea, under the name of Sosos, as mentioned by Berosus.*

It is stated by several writers, both Persian and Grecian, that, besides the Sosos of 60 years, the Chaldeans had in use several other cycles, one of 600 years, called the Neros, another of 3,600 years They had also a period called the Saros, consisting of 223 complete lunations in 19 years, after the expiry of which period, the new and full moons fall on the same days of the year.

What has been stated in support of the proposition that the astronomies of existing Eastern nations had a common origin may be summarised as follows:—(1) They had a like religious belief; (2) A like number of days of the week, with like names; (3) Similar divisions of the Ecliptic; (4) The same signs of the Zodiac; and (5) Similar months of the year. Also, (6) A like number of Lunar constellations; (7) A like use of the Celestial Sphere; (8) A like use of the Gnomon; (9) A like fantastical nomenclature of constellations; (10) Like ideas concerning Mythology; and (11) Simi-

* We read that Berosus, a native of Babylonia, and the High Priest of Belus, when the country was invaded by Alexander, became a great favourite with that Monarch, and wrote out for him a history of the Chaldeans. He mentions that there were accounts preserved at Babylon with the greatest care, which contained a history of the heavens, and of the sea and of the birth of mankind; that some time after the flood Babylon was a great resort of people of various nations who inhabited Chaldea, and lived without rule and order. Some fragments of this work, which was in three books, have been communicated by Eusebius.

lar cycles of 60 years; and, no doubt, other similarities might be traced.

Whatever controversies have arisen with regard to the details of differences or similarities between the systems of astronomy obtaining in various countries; whatever, also, may be the true facts as to the order in which each nation may have acquired its system, there is, at any rate, enough in those similarities to circumstantially establish, as a truth, the conjecture that the foundation of Prehistoric astronomy is to be found amongst those peoples of Central Asia who are generally referred to as the Aryan race.

CHAPTER II.

EARLY HINDU PERIODS.

In our endeavours to become acquainted with the earliest periods in which Hindu astronomy was extant, we are led into the Prehistoric age, which has, to some extent, been considered; an age comparable to early dawn, in which everything is still in a state of obscurity, the feeble twilight of those far-distant times enabling us only to perceive that there are objects around us which have a real existence, but the shadowy forms of which are extremely indistinct, and scarcely separable from the surrounding gloom. So far as can be traced, the basis of that science was in the religious aspirations of Hindu votaries, in times when each heavenly body represented a Divinity. The study of astronomy originated in the doctrine that the Supreme Being had assigned duties to each of the heavenly bodies, by which they became rulers of the affairs of the world, and that a knowledge of the Divine will would be acquired by watching and observing the order of their motions and the recurrence of times and seasons.

The early religion, indeed, of the Hindus, like other religions, had, as we know, a close intimacy with times and seasons; and there was in connection with their rites and ceremonies, a calendar to set forth the order in which they should be observed. This calendar, in the early periods referred to, had naturally an imperfect character, which led to methods afterwards adopted for its improvement, generally with a view to its adaptation to religious rather than to secular uses.

Now, among all nations the fundamental periods of time, the day, the month, the year, are the same, the variations occurring in them being principally in the arrangement of the days to form months and years; in the subdivisions of the day; in the times to be reckoned as the commencement of the day, whether at midnight, sunrise, or noon; in the subdivisions of the year into months, differing from each other as to the number of days of each; in the various kinds of months to form the year, and the like. Though there has apparently been the similarity to which allusion is made, nevertheless, there appears to have been in all nations a certain difficulty experienced as to the time when the year should be reckoned to begin, and in the consequent arrangement of the months and seasons, so that these should recur at regular intervals. With a view, therefore, to the establishment of some methodical data whereby to regulate these, people of all nations have had more or less necessity for observing with attention the motions of the heavenly bodies.

With the Hindus, this study became a sacred duty, at least amongst the more educated classes, inasmuch as the celestial bodies were viewed as Gods, and the worship of them was enjoined by the Vedas. Thus, the piety of the Hindus in primitive ages led them to watch with care all the phenomena of the heavens, and to perfect their calendar of festivals, etc., and to this end the first Hindu astronomers must have directed their particular attention. Their peculiar systems of algebra and arithmetic seem to show that these branches of science had their origin in the necessary requirements of their astronomy; and, indeed, so far as algebra is concerned, it is not improbable that this science was invented by them. At any rate, they attained to considerable proficiency in mathematics, as is clear from the methods employed by them to reconcile the motions of the sun and moon, so as to construct the period called the Luni-Solar year.

Amongst the injunctions enforced by the Institutes of Menu is contained a remarkable one making it imperative that the professions and trades pursued by the people should be followed only by those distinctively taught in them. Under the rule thus enjoined, each trade or calling came to be followed by distinct families, the secrets and artifices of such trade or calling being preserved in exclusive classes and sects of the population. The knowledge acquired by the Hindu astronomers was similarly guarded, with the greatest care, as sacred, and was supposed to be so secret that it was not known even to the Gods. It was not to be communicated to the common people, and, being regarded as a revelation to inspired saints, was only to be divulged to disciples similarly inspired.

This secretiveness has probably contributed, in some degree, to the difficulty now experienced in tracing the early history of the science of astronomy amongst the Hindus; for that part of it which was most ancient would no doubt be transmitted orally, and the science itself contained only in traditional statement.

At the end of the last century a great spirit of inquiry existed among our own countrymen in the East, and researches regarding Indian philosophy, literature, and science, were carried on with enthusiastic zeal and ardour, and with proportional success.

Among the subjects which were eagerly studied were those which related, for the most part, to the antiquity and ancient civilization of the Hindus.

It was then a general opinion (which, indeed, has existed both in ancient and in modern times) that the Hindu was one of the oldest of civilized nations, and it was sought to ascertain what ground there was for this opinion. Attempts were accordingly made to frame an authentic system of chronology, applicable to Hindu history. But, unhappily, the Hindus themselves have been long addicted to fabulous accounts of their own early history.

In their eyes, the present Kali age is one of degradation and misery, and their traditions lead them to magnify everything that relates to the past. They especially refer to the events that are supposed to have happened in the Golden and Silver ages, when (as they say) they were a free people, and when men were pure and free from disease; and to events of the Brazen age, when it is supposed

Yudhisthira and Rama, those heroes, whose glorious but fabled deeds are recorded in the great Epic Poems of India, must have lived and reigned.

Laying aside, as incredible, the accounts of their national existence for millions of years, given by the Hindus themselves, Sir W. Jones and Captain Wilford each investigated their records in the hope of finding authentic or probable dates for men who have lived, and events which undoubtedly must have occurred in past ages. Both, however, gave up the task as hopeless, though each furnished a table with a few probable dates, and Sir W. Jones, in conclusion, declares the subject to be so obscure and so much clouded by the fictions of the Brahmins that we can hope to obtain no system of Indian chronology, to which no objection can be made, unless the astronomical books in Sanscrit shall clearly ascertain the places of the colures in some precise years of the historical age, and not based on loose traditions like that of a coarse observation by Chiron, who, possibly, never existed. In a subsequent part of this work, an attempt will be made to establish some of the more important dates, so far as relates to matters connected with astronomy, by a reasoning based on the places of the colures, as well as by other means.

The Hindu writers are charged by their enemies with falsifying and exaggerating dates, a charge which appears to have been also made against other nations of great antiquity. The Egyptians, the Chinese, and the Persians, have each been accused of vanity in ascribing great antiquity to their several nations. The Chinese wish to pass themselves off as the oldest nation in existence. The Egyptians boasted to the Greeks that in their ancient writings they had accounts of events which happened forty-eight thousand years before, and the Babylonians also maintained that they had actual observations of astronomical phenomena made many thousands of years before. Calisthenes sent home to his uncle, Aristotle, copies of their observations, which were reported to have been made during the 19 centuries before that time (circ. 350 B.C.)

Now, all these were populous nations in times during which their history is regarded as authentic, and long periods must have elapsed in Prehistoric times before these people could have multiplied in population to the size they were when historians give first accounts of them. The apparent exaggeration in the descriptions which these nations give of their ancestors must be partly ascribed to romantic tales and traditions connected with such ancestors, mixed, in transmission, with events which actually happened.

One cause of the seeming exaggeration in chronology of remote times may have been our misapprehension of the different meanings ascribed to the term "year." To modern European nations it conveys only one meaning. What is termed a civil year was fixed in its present form by Pope Gregory, to remedy the inconveniences experienced by the various meanings then applied to the term. But the civil year, as we go back in history, was applied to periods of time very different from that which it defines now, namely, the time which the earth takes to complete its tropical revolution about the sun. As applied in remoter times, the term year has a less and less distinct meaning, until it loses its present character altogether. Various periods of time were in use, which historians have interpreted to signify years, such as our own. Some ancient astronomers gave the name of year to the times of revolution of each of the planets. Thus, Mars, Jupiter, and Saturn had each their years, consisting of the number of days they severally required to complete a revolution. The moon had its period of 30 days, and we read of different countries, in which the people have had a method of reckoning time, in periods of the moon, of two, three, or four months, being a much easier method than by the Solar year when great accuracy was not deemed necessary. This, of course, arose from the visible changes in its disc, from the day when it was not seen (at conjunction with the sun), to the day when it had a full round disc (in opposition to the sun); but in the Solar period of a year the day of the beginning could not be distinguished from other days.

The occupation of the people in countries where they were employed cultivating the ground or tending their flocks, suggested shorter periods than a year in reckoning the time for preparing the soil, for planting and sowing seed, for raising and harvesting crops, and for the time also when their domestic animals brought forth their young; all of which had some connection with a division of the year into three, and again into four, and sometimes into six seasons.

The ancient meaning of the word "year" would appear to be exceedingly ambiguous. Costard, in his "Rise of Astronomy," remarks that it was employed to denote any revolution of the celestial bodies, Solar, Lunar, or Planetary. In more recent periods it was applied to the apparent annual revolution of the sun. But, previously, the term has been applied to various periods of time. Thus, Plutarch, in the Life of Numa and Pliny, Lib. 7 Cap. 48, asserts that the Egyptian year was really a month, and, again, that four months was also used as the length of a year, which may probably have had its origin in an ancient division of the year into three seasons, a custom common to the Egyptians, the Greeks, and the Hindus. The Hindus, with a division of three—the dry, the rainy, and the cold—had also a division of six seasons in the year. It is argued that, as the Egyptians, so the Hindus, might anciently have computed by periods of two or of four months. Hence the exaggeration in their chronology.

It is further suggested that the Children of Israel, during their captivity in Egypt and long afterwards, may have followed the Egyptians in this mode of reckoning the length of the year; and that the supposed exaggeration of the patriarchal lives may have been reckoned by years of this kind, which would bring them down to the ordinary length of the lives of men of the present age.

So, also, the 48,000 years during which the Egyptians said they had records, reckoned by years of Lunar periods, may have been only 4,000 of our years.

We may speak with a greater degree of certainty of events that have happened in times when the meaning of the year approximated even nearly to that which is given to it in these days; but it is impossible to form a conjecture regarding the absolute period when events are said to have occurred if we do not know the meaning to be applied to the word "year."*

Of the great antiquity, however, not only of the Hindu nation, but of Indian astronomy generally, the first evidence was afforded to European investigators by the publication of certain astronomical tables, in the "Memoirs of the Academy of Sciences," in 1687, which were brought from Siam by M. Le Loubère, of the French Embassy, and subsequently examined and explained by the celebrated Cassini. These were, and are, known as the "Tables of Siam," Two other sets of tables were afterwards received from French missionaries then in India. These are called the "Tables of Chris-

---

* "According to Pliny the Chaldeans boasted that they had a regular series of astronomical observations engraved upon bricks for the space of 720 thousand years, but it was afterwards proved by Dr. Jackson, in a long series of quotations, that this calculation by years should have been days, and that Abydenus, who copied the public records kept at Babylon, improperly interpreted the word Jomin—signifying days—in the sense of years; which interpretation that term, as well as the Hebrew word Jamin, will also bear."—*Dr. Jackson's Chronology, Antiq. Vol. I., p.* 200.

According to Diodorus Siculus, lib. I., and Varro, quoted by Lactantius (de Origine Erroris, lib. 2—sec. 12), the Egyptians in the most early days computed time by a lunar year of 30 days.

According to Pomponius the Egyptians boasted that during the immense period of the existence of their empire the stars had four times changed their course, and that the sun had set twice in the quarter in which he now rises.

QUERY: Did the inventors of this fable coast round Africa, sailing down the Red Sea and enter the mouth again and sail up the Nile?

nabouram and Narsapur"; but they remained unnoticed till the return of the French astronomer, Le Gentil, who had been in India for the purpose of observing the Transit of Venus in 1769. During his stay there, he employed himself in acquiring a knowledge of Indian astronomy, being instructed by the Brahmins of Tirvalore in the method used by them in calculating eclipses; and they communicated to him their tables and rules, which were published by Le Gentil as the "Tables of Tirvalore," in the Memoirs of the Academy of 1772.

It is, however, to another Frenchman, M. Bailly, the author of "Traite de l'Astronomie Indienne," that we owe the full discussion regarding the antiquity of the four tables above referred to, to which he devoted an entire volume.

Professor Playfair made an elaborate investigation of Bailly's work, and presented it to the Royal Society of Edinburgh, in a long paper, which was published in their transactions in 1790.

In introducing the subject, he says:—"The fact is that, notwithstanding the most profound respect for the learning and abilities of the author of 'Astronomie Indienne,' I entered on the study of that work not without a portion of the scepticism which whatever is new and extraordinary ought to excite, and set about verifying the calculations and examining the reasons in it, with the most scrupulous attention. The result was an entire conviction of the one, and of the solidity of the other."

By elaborate calculations, founded upon the best modern tables of the time (those of Lacaille and Mayer) and by going back to the epoch of the tables of Tirvalore, which was midnight between the 17th and 18th February, 3102 B.C. (at which time the sun was entering the Moveablo Zodiac, and was in Long. 10 signs 6 degrees), and by computing backwards the places of each of the bodies (the sun and the moon), an exact agreement was found to exist between such places at the epoch mentioned and the places given by the tables of Tirvalore.

The general conclusions established from a comparison of these calculations were as follow:—

I.—That the observations on which the astronomy of India is founded were "made more than three thousand years before the Christian Era, and, in particular, the places of the sun and moon were determined by actual observation."

II.—That, "though the astronomy of the Brahmins is so ancient in its origin, it contains many rules and tables that are of later construction."

III.—That "the basis of the four systems of astronomical tables, which have been examined, is evidently the same."

IV.—That "the construction of these tables implies a great knowledge of geometry, arithmetic, and even of the theoretical parts of astronomy."

The opinion of Bailly, however, that a general conjunction of the sun, moon, and planets at the time stated (3102 B.C.), was known to the Hindus from actual observations was much controverted at the end of the last century.

That there was an approach to such a conjunction was generally admitted, yet it was only an approach.

Consequently, an argument against Bailly's opinion was advanced to the effect that the epoch of 3102 B.C. was adopted by the Hindus at a comparatively recent date, only from calculation. For a further discussion in respect to this controversy, the reader is referred to Appendix I.

It may be here mentioned that, in the course of Bailly's investigation of the tables from Chrisnabouram, he had observed in the correction given in these tables, for finding the true place of a planet from the mean, "that the magnitude of it was applied with no small exactness, and that it varied in different points of the orbit by a law which approached very nearly to the truth." What, then, was the method employed by the Hindus in making their calculation of the correction?

D

Cassini had previously found that the equation in which this occurred—which, with us, goes by the name of "the equation of the centre"—followed the ratio of the sines of the mean distance from the apogee, but it was calculated only for a few points in the tables of Siam, and it could not be ascertained with what degree of accuracy the law was fully observed. From the tables of Chrisnabouram, however, Builly found that the law was nearly observed, but only nearly. On this he concluded that this law of the sines was not the one which was followed or intended to be followed in the calculation.

Playfair, then, endeavoured to reconcile these irregularities with a theory of his own. He assumed a double eccentricity for the orbit, and from this hypothesis he deduced a formula, which agreed well with the corrections given in the tables.

Now, in the subsequent parts of this work an endeavour is made to show that Playfair's assumption was not the real hypothesis, but that, unlike the Epicycles of Ptolemy, the Indian Epicycles had a variable circumference, that of the first Epicycle being largest at Apogee and Perigee, varying from those points through the deferent to its places at the quadrants, where its circumferences were least.

In the cases of Mercury, Venus, and Mars there was the same kind of variation, but in those of Jupiter and Saturn the greatest circumference was at 90 degrees from the line of Apsides.

Whatever may be the truth as to the origin of the interesting tables which have given rise to so much discussion, it is certain that the ancient Hindu astronomers, many centuries before the Christian Era, were in possession of knowledge, derived from observations made by them of the motions of the heavenly bodies, which they were able to use, and did actually use, in very accurate computations of time. It is also abundantly clear from writings of the Hindu astronomers of later date, which refer to those earlier astronomers and their traditioned observations, that the latter were well acquainted with the nature of the phenomenon of the precession of the Equinoctial point, and in their computations, arrived at its annual rate with a considerable degree of accuracy.

Of course, the Hindus, as we ourselves, were compelled to assume some epoch at which the motion of a heavenly body might be supposed to begin. The opinion expressed by Laplace that the epoch of the Kali Yuga (3102 B.C.) was invented for the purpose of giving a common origin to all the motions of the planets in the Zodiac, is no doubt very true; but the beginning of the Kali Yuga would appear to be only one of several epochs, at which, according to the Hindu astronomy, there was a conjunction of the sun, the moon, and the planets.

For instance, the Hindus had certain assumed epochs, carrying the mind back to dates when the heavenly bodies were supposed to be in conjunction, and from whence their motions were presumed to commence—in short, to the period of creation—and even beyond such a period, *e.g.*, to the beginning of a day of Brahma, which day they called a Kalpa. Even this Kalpa was only a part of Brahma's life.

The Kalpa was a period of 4,320,000,000 years. One-thousandth part of this was the Maha Yuga, or Great-Yuga.

The Maha-Yuga again was further subdivided and made up of the Kali, the Dwapara, the Trita, and the Krita Yugas, thus:—

The Kali Yuga (one-tenth of Maha-Yuga) ..	=	432,000	
„ Dwapara (twice the Kali)	=	864,000	
„ Trita (thrice the Kali)	=	1,296,000	
„ Krita (four times the Kali)	=	1,728,000	
The Maha-Yuga = Sum	=	4,320,000	
„ Kalpa	=	4,320,000,000	years.

At each of these commencing epochs Hindu astronomers considered that the moveable celestial bodies were in conjunction.

Thus, it will be seen that any one of the above epochs might be used for the purpose of computing the mean places of each; and as

D 2

the Kali, the smallest period of all, was just as useful as the others for this purpose, it alone was generally used, and was, as before shown, supposed to begin at midnight between the 17th and 18th February, 3102 B.C. So that this epoch is one of great importance in considering problems affecting Hindu astronomy as well as questions relating to their civil time.

The Kalpa and its subdivisions, although appearing at the first blush so ponderous and ridiculous, will be shown in a subsequent part of this work to have been really useful in computations of various astronomical problems for the purpose of reducing errors to a minimum and of ensuring accuracy. In short, the Hindus used these great assumed periods much in the same way as we use decimal fractions to eight or nine places when expressing elements relating to the planets (the decimal system not being then known).

A difference of opinion existed among the more ancient astronomers as to whether their calculations ought to begin from the beginning of Brahma's life or the beginning of a Kalpa; and it is suggested in the Surya Siddhanta that the end of the Krita Yuga is a convenient epoch, from which to compute easily the terrestrial days, and to find the mean places of the planets.—thus:—

"For at the end of this Krita Yuga, the mean places of all the planets, except their nodes and apogees, coincide with each other in the first point of Mesha [or Stellar Aries], then the place of the moon's apogee is nine signs, her ascending node is six signs, and the places of the other slow-moving apogees and nodes, whose revolutions are mentioned before, are not without degrees [*i.e.*, they may have some degrees of longitude]."

It was deemed by writers in other Siddhantas that any epoch deduced from the rules, if it agreed with the observed position of the planet, might be assumed. Davis says that modern Hindu astronomers do not go further back than to some assigned date of the Saca (A.D. 78), when, having determined the planets' places for that time, they compute the mean places for any other time by means of tables, etc.

CHAPTER III.

THE HINDU ECLIPTIC.

In the astronomical systems of nearly all Eastern nations there existed, at the earliest historical dates, an intelligent grasp of the apparent motions of the sun and the moon, in their respective paths in the Celestial Sphere. Connected, as these were, with the religious observances of the Hindus and other nations, not only did their periodic revolutions give rise to the construction of calendars, but when those heavenly bodies or the planets were eclipsed or occultated, such phenomena undoubtedly originated important calculations as to the periods of their recurrence.

Hence, even in anti-historic times, the nature of the Ecliptic was well understood, and, at the earliest known periods, Asiatic astronomers, as has been suggested in a previous chapter, divided the Ecliptic and the Zodiac into 28 parts, forming so many groups of stars in the path of the moon, each division corresponding nearly with the space of the moon's daily motion through them. The groups were hence called the Lunar Asterisms.

The systems of most Eastern countries generally resembled each other in formation, differing only in minor particulars, such as the extent of each constellation, the number of stars included in it, or as to which was regarded the principal star of the cluster.

A comparison of the system of Lunar Asterisms existing in the astronomy of Eastern countries with those of Western nations has, to some extent, been discussed in Chapter I., and it is not intended to follow the various writers who have made this comparison the subject of discussion, except in so far as they may refer to the Hindu system.

At some later period than that of the Hindu Aryan migration, and yet antecedent to any historical record, the Hindu astronomers improved upon their system of Lunar Asterisms, by reducing the number of divisions of the Ecliptic from 28 to 27, and by making them all equal to one another, so that each should extend over 13° 20′ or 800 minutes of arc on the Ecliptic, by which means the constellations were made to agree more nearly with the moon's mean daily motion. As the actual time for a mean sidereal revolution in 27.3216 days, 27 was the nearest whole number of days suitable for the division of the Ecliptic. Moreover, it was a more convenient number than 28, for calculation, in reducing all their observations to a system.

The Hindus, unlike the ancient Chinese, had not the ambition of making a catalogue of all the stars which were visible to them. They had a more important object in view, namely, the study of the motions of the sun, the moon, and the planets, and other astronomical phenomena, primarily for the purpose of computing time, and of constructing and perfecting their calendars. Such an object, they knew, could not be materially advanced by ascertaining merely the positions of stars fixed beyond, or outside the course of the moving celestial bodies; and they accordingly confined their attention to those stars which lay in the moon's path, immediately North or South of the Ecliptic — stars which are liable to be occultated by the moon, or which might occasionally be in conjunction with it and with the planets.

By thus confining attention to the stellar spaces in the vicinity of the Ecliptic, their system was rendered, in the main, independent of the use of astronomical instruments, and dependent mostly on calculation for the accuracy of their observations.

Hence the ancient Hindu astronomers chose a set of 27 principal stars, one for each of the 27 Lunar Constellations, in general the brightest star of the Asterism, and called it Yoga-tara, whilst the Asterism cluster was named the Nacshatra. The Yoga-tara was connected with the beginning, or first point, on the Ecliptic of the division representing the space of the Asterism by the small arc of apparent difference of longitude between them, this arc being called the Bhoga of the Asterism.

Plate V.

THE HINDU ECLIPTIC
AND NORTHERN YOGATARAS WITH MODERN NAMES.

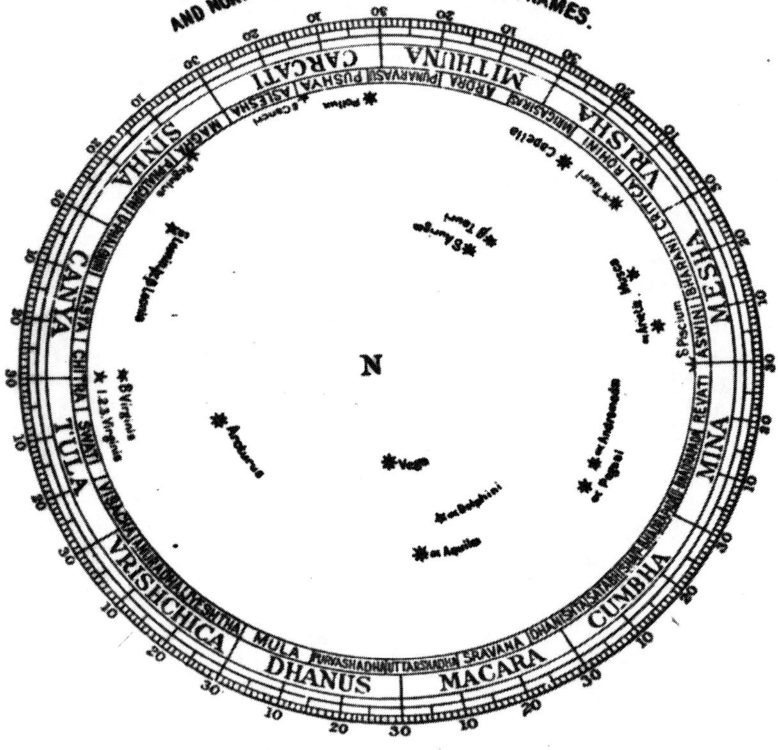

Plate VI.

THE HINDU ECLIPTIC
AND SOUTHERN YOGATARAS WITH MODERN NAMES.

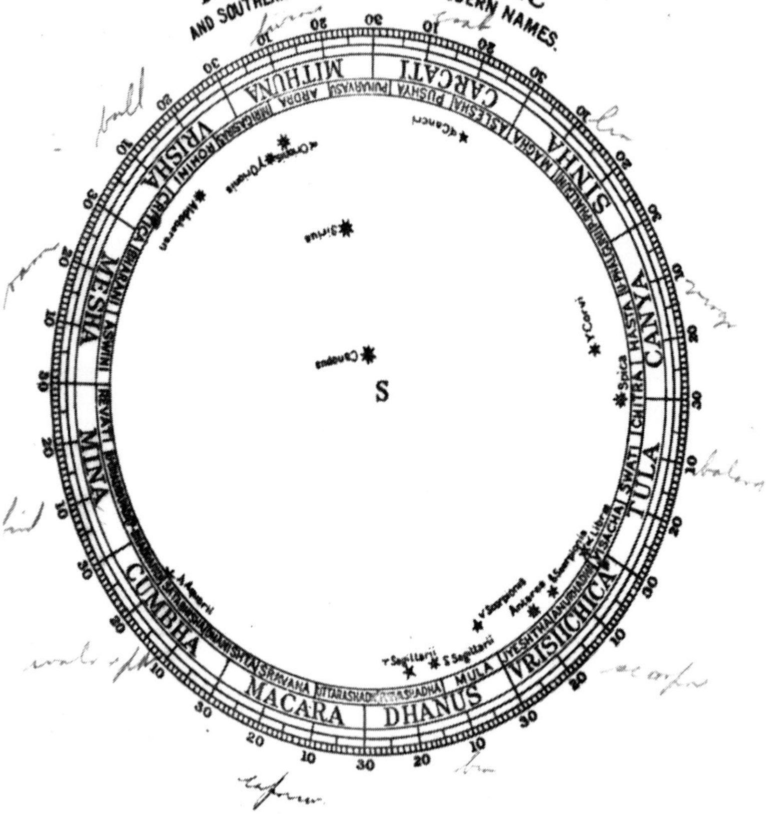

Thus, the 27 divisions of the Ecliptic became as fixed in position as the stars themselves, like a great fixed dial, with the numbers ranging, not along the Equator, but along the Ecliptic itself.

The accompanying diagram (Fig. IV.) may, perhaps, more

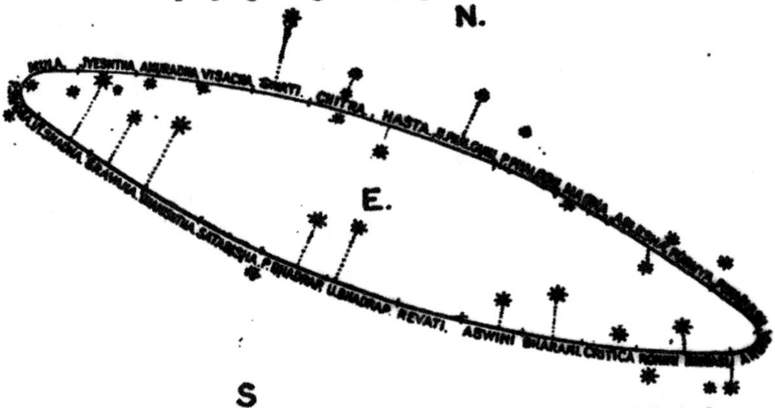

explicitly convey the nature of the Hindu Ecliptic, which is here shown, as a great circle in perspective.

Each division represents one twenty-seventh part of the Ecliptic, and each star the Yoga-tara of the Nacshatra, or Lunar Asterism, to which it belongs.

It will be observed that the Yoga-tara might be either in the Northern or the Southern Hemisphere, and the stars selected were those most suitable for observation, either on the Ecliptic or near it, North or South, but always such as were capable of being occultated by the moon or of being in conjunction with it or with the planets.

To render this important part of Indian astronomy still more easily understood, the two accompanying Plates, V. and VI., are intended as a graphic representation of the Hindu Ecliptic, and of the Lunar Asterisms, together with the Solar signs of the Hindu Zodiac, the position of each being fixed by a supposed projection of the Yoga-tara on the plane of the Ecliptic, the Northern stars, with their modern names, on one side of the plane, and the Southern stars on the other, the divisions retaining the same names in each Hemisphere.

In addition to the Yoga-taras, or principal stars, the Siddhantas give the names and places of a few other stars, among the most important of them being the Southern stars, Agastya, or Canopus, and Lubdaca (the Hunter) or Sirius, and the Northern stars Bramehridya or Capella, and Agni or β Tauri, Prajapati δ Aurigæ.

Of the few stars which were close to the Ecliptic, and most suitable as points of reference, Magha, Pushya, Revati, and Sataraka (or, respectively, Regulus, δ Cancri, ζ Piscium, and λ Aquarii) are recommended as most to be preferred for planets whose longitudes are required, or any star of Chitra (as α Virginis) having very inconsiderable latitude, when the planet, having also nearly the same latitude, appears to touch the star. The stars at a greater distance from the Ecliptic would only occasionally be in conjunction with the planets or be occultated by the moon, giving their own fixed longitudes and latitudes to the moveable bodies with which they were for the moment coinciding. By a little previous calculation regarding the stars with which particular planets and the moon would be in conjunction, the times of occurrence were easily ascertained, and, with care, even by unaided vision, ordinary observations of a fair degree of accuracy would be obtained.

The Asterisms are described in three stanzas of the Retnamala; they are :—

Lunar Constellation figured by

1. Aswini—A horse's head.
2. Bharani—Yoni or Bhaga.
3. Critica—A razor.
4. Rohini—A wheel carriage.
5. Mrigasiras—The head of an antelope.
6. Ardra—A gem.
7. Punarvasu—A house.
8. Pushya—An arrow.
9. Aslesha—A wheel.
10. Magha—Another house.
11. Purva Phalguni—A bedstead.
12. Uttara Phalguni—Another bedstead.
13. Hasta—A hand.
14. Chitra—A pearl.
15. Swati—A piece of coral.
16. Visacha—A festoon of leaves.

Lunar Constellation figured by

17. Anuradha—An oblation to the gods.
18. Jyeshtha—A rich ear-ring.
19. Mula—The tail of a fierce lion.
20. Purvashadha—A couch.
21. Uttarashadha—The tooth of a wanton elephant, near which is the kernel of the sringataca nut.
22. Sravana—The three-footed step of Vishnu.
23. Dhanishta—A tabor.
24. Satabhisha—A circular jewel.
25. Purva Bhadrapada—A two-faced image.
26. Uttarabhadrapada — Another couch.
27. Revati—A smaller sort of tabor.

PLATE VII.

THE HINDOO LUNAR MANSIONS.

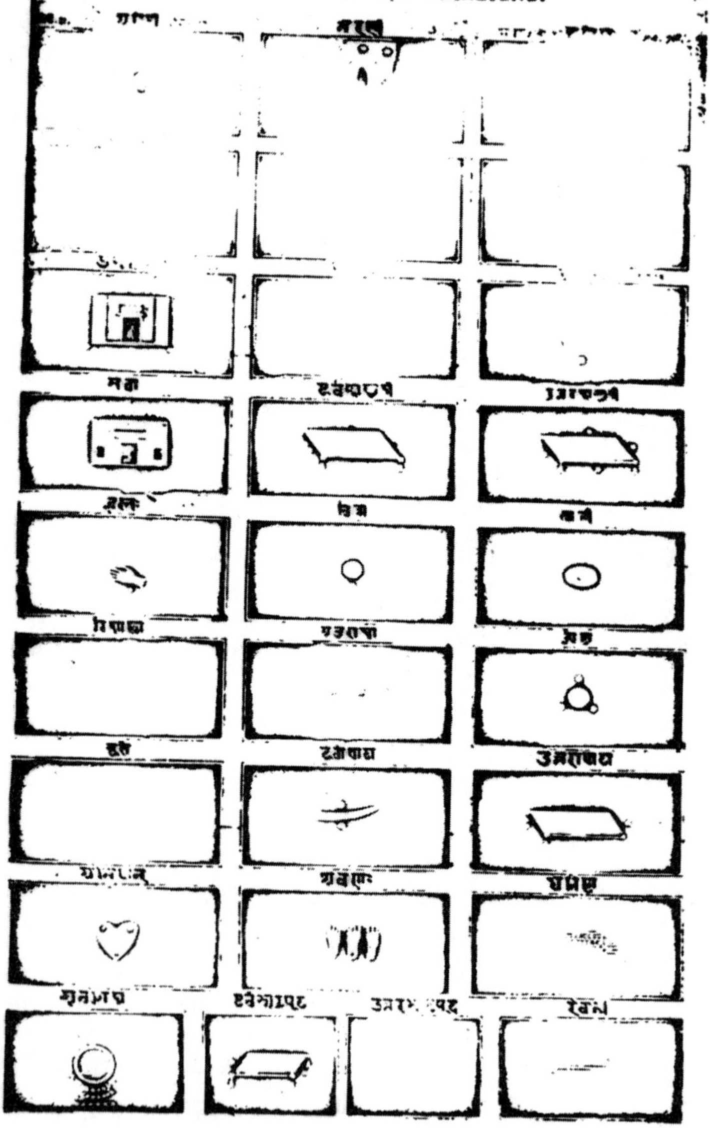

The symbols by which the Indian Nacshatras were represented, as given by Sir William Jones, are shown in the accompanying plate. [Plate VII.]

In pursuit of his enquiry regarding the particular stars which gave names to the Indian divisions of the Zodiac, Colebrooke formed a list, founded upon the several Indian works of astronomy, the Surya and Brahma Siddhantas, the Siddhanta Siromani, the Grahalagava, and the Siddhanta Sarvabhanma, in which the apparent latitudes and longitudes of the principal stars of each of the 27 constellations were laid down with a greater or less degree of accuracy. In this list the modern name is given of that star which is supposed to be identical with the Yoga-tara of each Asterism, and as agreeing most nearly, in position, with that which is generally assigned to it in the Siddhantas.

Bentley gave a similar list in his "Hindu Astronomy," and another one is given in Burgess's translation of the Surya Siddhanta.

The accompanying table is taken principally from Colebrooke, whose orthography has been retained, being the same as that of Sir W. Jones and Davis. The number of stars comprised in each small group are those given by the Hindu astronomer, Sripati, but they are far from being all that were known of those constituting the Lunar Mansions.

The apparent latitudes and longitudes are those deduced from the Bhogas of the Surya Siddhanta, a corresponding list from Brahmegupta being given by Bhascara, in which the latitudes and longitudes of the Yoga-taras differ in five of them, the difference in longitude being 1°, 3°, 3°, 2°, and 4° respectively.

These differences may have been partly owing to mistakes in the selection of the stars intended as the Yoga-taras, and partly to the greater difficulty of observing stars that were at a great distance from the Ecliptic. Thus, such stars as Vega, the principal star of Abhijit, and Arcturus, the bright star of Bootes, being so far removed from the paths of the sun, the moon, and the planets, would be seldom objects for observation.

The stars selected for fixing the points of division of the Ecliptic are apparently those best adapted for marking out the path of the moon in the heavens.

Number of Asters along Ecliptic.	NAME.	Number of Stars in Asters according to Sripati.	Apparent longitude from Surya Siddhanta.	Apparent latitude from Surya Siddhanta.	Principal Star of Asterism according to Colebrooke.	Principal Star according to Burgess.	Principal Star according to Bentley.
1	Aswini	3	8° 0′	10° 0′ N	α Arietis ..	β Arietis	γ or β Arietis.
2	Bharani	3	20 0	12 0 N	Musca	35, 41 Arietis.. ..	35 Arietis.
3	Krittika	6	37 30	5 0 N	π Tauri ..	Alcyone 27 α 28 Tauri	Alcyone.
4	Rohini	5	49 30	5 0 S	α Tauri ..	Aldebaran	Aldebaran.
5	Mriga	3	63 0	10 0 S	λ Orionis ..	λ Orionis	113, 116, 117 Tauri.
6	Ardra	1	67 30	9 0 S	α Orionis ..	α Orionis 135 Tauri ..	133 Tauri.
7	Punarvasu	4	93 0	6 0 N	β Geminorum..	Pollux	Pollux.
8	Pushya	3	106 0	0 0 N	δ Cancri ..	δ Cancri	δ Cancri.
9	Aslesha	5	109 0	7 0 S	α 1 and 2 Cancri	ε Hydra and α Cancri	49, 50 Cancri.
10	Magha	5	129 0	0 0 N	α Leonis ..	Regulus	Regulus.
11	Purva-Phalguni .	2	144 0	12 0 N	δ Leonis ..	δ Leonis	70, 71 Leonis.
12	Uttara-Phalguni ..	2	156 0	13 0 N	β Leonis ..	β Leonis	β Leonis.
13	Hasta	5	170 0	11 0 S	γ or δ Corvi ..	γ and δ Corvi	7, 8 Corvi.
14	Chitra	1	180 0	2 0 S	α Virginis ..	Spica	Spica.
15	Swati	1	199 0	37 0 N	α Bootes ..	Arcturus	Arcturus.
16	Visakha	4	213 0	1 30 S	α or χ Libræ ..	ι or χ Libræ	24 Libræ.
17	Anuradha	4	224 0	3 0 S	δ Scorpionis ..	δ Scorpionis	β Scorpionis.
18	Jyestha	3	229 0	4 0 S	α Scorpionis ..	Antares	Antares.
19	Mula	11	241 0	9 0 S	λ Scorpionis ..	λ Scorpionis	34, 35 Scorpionis.
20	Purva-Shadha ..	2	254 0	5 30 S	δ Sagittarii ..	δ Sagittarii	δ Sagittarii.
21	Uttara-Shadha ..	2	260 0	5 0 S	τ Sagittarii ..	τ Sagittarii	φ Sagittarii.
22	Abhijit	3	266 40	60 0 N	α Lyræ ..	Vega	Vega.
23	Sravana	3	280 0	30 0 N	α Aquilæ ..	α Aquilæ	α Aquilæ.
24	Dhanishtha	4	290 0	36 0 N	α Delphini ..	α Delphini	β Delphini.
25	Satataraka	100	320 0	0 30 S	λ Aquarii ..	λ Aquarii	λ Aquarii.
26	Purva-Bhadrapada ..	2	326 0	24 0 N	α Pegasi ..	α Pegasi	α Pegasi.
27	Uttara-Bhadrapada ..	2	333 0	26 0 N	α Andromedæ	γ Pegasi or .. α Andromedæ ..	γ Pegasi.
28	Revati	12	359 50	0 0	ζ Piscium ..	ζ Piscium	ζ Piscium.

The Orthography of the names of the Asterisms is the same as in each of the three lists given by Sir W. Jones, Colebrooke and Davies.

At certain times of the year the beautiful clear Indian sky, visible all round from the house-tops, as a great hemisphere, is peculiarly favourable for astronomical observations; and the ancient astronomer, seated on his chunam terrace in the pleasant cool evenings, had

little need of astronomical instruments while patiently watching the moon and the planets in their course through the Zodiacal stars.

The well-known Yoga-taras among the fixed stars, and which the planets pass on their way, form so many immovable points, and like milestones on a road, furnish him with his means of observation. The relative times of passing such points suggested methods of calculation somewhat similar to those employed by ourselves, in solving simple questions, such, for instance, as the determination of the time when two hands of a clock in conjunction will be together again after any number of revolutions of either of them, or when we seek for the synodic periods of the planets, the times of new and full moon, and other problems of a like nature, data for the solution of which were well known in India many centuries before they were known in Europe; such problems formed the constant subject matter of the algebra of the Hindus, as contained in their astronomical works of the first centuries of the Christian Era.

Again, there are certain phenomena which occur regularly, near the Ecliptic, too remarkable not to have been observed by such patient observers as the Hindu astronomers.

For example, if we consider the facts relating to eclipses of stars by the moon in her course, the moon's node has a retrograde motion of about 3¼ mins. of arc daily, and in her progress she passes over all stars situated on the Ecliptic, and a star such as Regulus, whose latitude is only 27½ mins. of arc, must always be eclipsed when it is passed by the moon in her node. Such, also, would be the case with other stars on, or near, the Ecliptic. Moreover, these phenomena would be repeated continually in sidereal periods of 18⅔ years.

The Surya Siddhanta makes the sidereal period of the moon's node 6794.443 mean Solar days, whereas, according to our modern astronomy, it is 6793.39108 days.

It is upon the position of the moon's node at the time of conjunction or opposition of the sun and moon, that a Solar or Lunar eclipse depends; and if these bodies and the earth are all three in one straight line, an eclipse must happen, and the same eclipse will return in 6585.78 days, on in 18 years and 10 or 11 days, according as five or four leap-years occur during that time. Or in 18 Julian years 11 days 7 hours 43 minutes.

The same observations apply to all other eclipses which happen when the moon is near her node, within what are called the Lunar Ecliptic Limits. These all return after periods of the same length, so that a complete list of eclipses that occur in one such period will be sufficient for forming a list extending over several centuries, either past or future.

The Hindus were at a very early date well acquainted with these facts relating to eclipses. They had rules for calculation of the various phases both of Lunar and Solar eclipses, the times of beginning, middle, and end, as set forth in their various astronomical works, but they depended chiefly on those of the Surya Siddhanta.

Amongst the superstitious of all ancient nations we find that eclipses of the sun and moon had a terrible import, being supposed to presage dreadful events.

By the common people of the Romans, as also by the Hindus, a great noise was usually set up with brazen instruments, and loud shouts during eclipses of the moon. The Chinese, like the Hindus, supposed eclipses to be occasioned by great dragons on the point of devouring the sun and moon, and it was thought by the ignorant that the monsters, terrified by the noise of the drums and brass vessels, let go their prey.

The cause, however, of eclipses, notwithstanding the superstition of the people generally, was well understood by the Hindu astronomers, as is shown by the following extracts taken from the Siddhanta Siromani:—

"The moon, moving like a cloud in a lower sphere, overtakes the sun, hence it arises that the Western side of the sun's disc is first obscured, and that the Eastern side is the last part relieved from the moon's dark body; and to some places the sun is eclipsed, and to others he is not eclipsed."—(Siddhanta Siromani, ch. viii., par. 1).

"At the change of the moon, it often happens that an observer placed at the centre of the earth, would find the sun, when far from the Zenith, obscured by the intervening body of the moon; whilst another observer on the surface of the earth will not, at the same time, find him to be so obscured, as the moon will appear to him to be depressed from the line of vision extending from his eye to the sun. Hence arises the necessity for the correction of parallax in celestial longitude, and parallax in latitude in Solar eclipses, in consequence of the difference of the distances of the sun and moon. (id., par. 2).

"When the sun and moon are in opposition, the earth's shadow envelopes the moon in darkness. As the moon is actually enveloped in darkness its eclipse is equally seen by every one on the earth's surface, and as the earth's shadow and the moon which enters it are at the same distance from the earth, there is, therefore, no call for the correction of the parallax in a Lunar eclipse. (id., par. 3).

"As the moon moving eastward enters the dark shadow of the earth, therefore its Eastern side is first of all involved in obscurity, and its Western is the last portion of its disc which emerges from darkness, as it advances in its course. (id., par. 4).

"As the sun is a body of vast size, and the earth insignificantly small in comparison, the shadow made by the sun from the earth is, therefore, of a conical form, terminating in a sharp point. It extends to a distance considerably beyond that of the moon's orbit. (id., par. 5).

"The length of the earth's shadow and its breadth at the part traversed by the moon may be easily found by proportion. (id.. par. 6)."

A similar explanation is given in the Surya Siddhanta, but

the subject of eclipses was of a nature too sacred to be treated lightly or communicated indiscriminately, and a warning is delivered in the Surya Siddhanta to this effect:—"(Oh, Maya) this science secret even to the Gods is not to be given to anybody but to the well-examined pupil who has attended one whole year."

The Lunar Asterisms, contained in the divisions of the Indian Ecliptic, necessarily follow each other in the order of their principal stars, and, as in every other system, some point was necessarily selected upon the Ecliptic to mark the beginning of the system. The point at the present time marking such beginning was not the one which, in the ancient Hindu astronomy, was taken as the beginning. The fixing of the first point of the Indian Zodiac so as to make it unchanged in after time is characterised as a remarkable event in the modern Hindu astronomy, and, indeed, in this fixture is contained one of the fundamental differences between the Hindu and European systems. In the European astronomy, all longitudes are measured by arcs of the Ecliptic, whose origin is the Equinoctial point at the time of observation. This point, the origin of our longitudes, moves backwards along the Ecliptic at an annual rate of about 50" causing an annual increase of the same amount in the longitudes of all the stars, and this movement is in our phraseology termed the precession of the Equinoctial point. Thus, in our system, the origin of longitudes is perpetually changing. The Indian astronomers, however, avoid the annual change of longitudes by assuming a fixed point of the Ecliptic as the beginning of their system, the position of the Lunar Asterisms being all fixed in relation to that beginning. Not that they were ignorant of the precession. Indeed, the Hindu astronomers were, at a very early date, well acquainted with that phenomenon. It may here be desirable, perhaps, to state for the benefit of the general reader, what is understood by "the precession."

In the figure let C represent the centre of the earth, A ♈ D B the Ecliptic, π and π′ its Poles, E ♈ R Q the Equator, and P P′ its

Poles. A straight line, P C P′, will then represent the axis, about which the celestial bodies appear to revolve daily, from East to West, but about which the earth actually rotates from West to East, causing the apparent revolution of the stars. Let a great circle, ϖ P ϖ′ P′ be supposed to pass through the four Poles; this circle is called the Solstitial Colure; it bisects both the Ecliptic and the Equator, at right angles in the points A, B and E, Q. The point B is called the Summer Solstice, being the point of the Ecliptic nearest the Pole P, and the Winter Solstice is the point A. The line ♈ C ♎, in which the plane of the Equator intersects that

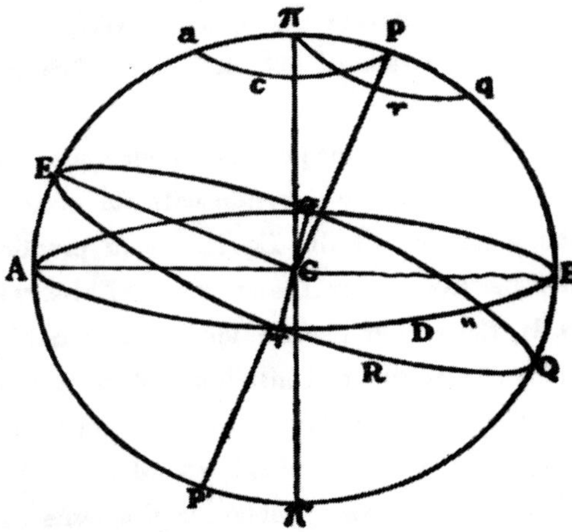

of the Ecliptic, meets the Celestial Sphere in the two points ♈ (or Aries) and ♎ (or Libra). The angle between the two planes, measured by the angle A C E or the arc A E, is called the Obliquity, and is equal to the angle ϖ C P, or the arc ϖ P. This Obliquity is now about 23° 27′, being formerly nearly 24°.

Let the small circle ϖ *r q* be supposed to represent the Arctic Circle of the Celestial Sphere; then, on account of the apparent motion of the Sphere about the axis C P, the point ϖ (the Pole of the Ecliptic) appears to move daily round the Pole P (the Pole of the Equator) in the circle *q r* ϖ; the apparent motion being caused by the rotation of the earth round its axis. The point ϖ as far as is known is a fixed point among the fixed stars, but the point P, the Pole of the Equator, though it seems to be fixed, near to the star α Ursæ Minoris (thence called the Pole Star) is, in fact, moving slowly from it with an imperceptible motion, in the small circle P *c a*, a circle which is parallel to the Ecliptic at a distance from π of 23° 27′ nearly. To complete a revolution in this circle about π, P's progress occupies nearly 25,780 years.

During the same period the line of intersection of the Ecliptic ♈ C ♎ and its two extremities, the Equinoctial points, and the two points A and B, where the Solstitial Colure meets the plane of the Ecliptic, are all moving slowly in that plane, and severally complete one whole revolution, at the rate of 50″ annually. This movement is called the precession.

Now, just as it has been of the greatest importance in the observations of modern astronomers to ascertain with the utmost exactitude the rate of the precession of the Equinoxes (a practical problem depending for its solution upon the accuracy of observations made at long intervals, the earlier observations being partly vitiated by the imperfections of astronomical instruments), so it was an important problem with the ancient Hindu astronomers to ascertain the same rate of regression, but they calculated with the aid of what we call the Solstitial Colure. The problem is the same as with ourselves, for the Solstices being at 90 degrees from the Equinoxes, the motions must obviously be uniformly the same, and it is only a matter affecting the convenience of the observer which method he will adopt.

It is distinctly stated by the commentators of the Surya Siddhanta (perhaps the most important extant astronomical work in the possession of the Hindus) that the apparent latitudes and longitudes of the stars given in it are adapted to that particular time when the Vernal Equinox did not differ from the origin of the Ecliptic in the beginning of Mesha. The origin here referred to is the fixed beginning point on the Ecliptic of the 12 Indian Signs of the Zodiac, and it is also the beginning of the 27 Nacshatras, or the first point in Aswini. [See Plates V. and VI.]

It has been further ascertained that the origin referred to is a point of the Ecliptic ten minutes East of the star Revati, (or ζ Piscium).

If, then, as undoubtedly is the case, the origin (or commencement), for present computations, on the Hindu Ecliptic, was fixed at the first point of Aswini, it is of importance to ascertain the date when this fixture, or change of system, took place; because all ancient Hindu astronomical dates are, in a measure, dependent for their accuracy on a knowledge of it.

As will presently appear, the date is a comparatively recent one, but, nevertheless, when established, it affords a means of tracing earlier dates, which, without it, could not readily be authenticated.

Several methods, then, have been employed for the purpose of ascertaining the period when the first point in Aswini was, in Hindu astronomy, established as their origin of apparent longitudes; thus:

(1) According to a statement of the celebrated Indian astronomer, Brahmagupta (as well as of other astronomers), the star Revati (or ζ Piscium) had no longitude or latitude in his time, which implies that it was then in the Equinoctial point.

To calculate the date at which this occurred, *i.e.*, the time when Brahmagupta lived, Colebrooke found from Zach's tables the right ascension of the stars in A.D. 1800, to be 15° 49′ 15″ and the precession in Right Ascension being reckoned at 46′ 63″; from this, as from a mean value, he estimated that the elapsed time for the regression of the Equinoctial point from its position when it coincided with the star ζ Piscium, and its position in 1800 A.D., would amount to 1221 years, which, taken from 1800, gave 579 A.D. as the date of Brahmagupta's assertion, and, approximately, the date when the first point of the Asterism Aswini was made the origin of Indian longitudes.

(2) Bentley, from a comparison of the longitude of Regulus, as recorded by Indian astronomers, with the longitude of the same star as given in the tables of the British Catalogue of 1750, and computing with a mean rate of regression in longitude of 48·5666″, and allowing a secular variation of 2·27″ for every century (being the diminishing rate as we go back into antiquity), estimated the date when the Equinoctial point and the beginning of Aswini were coincident to be 538 A.D.

(3) Again, in a note to an article (I., 27) of Burgess's translation of the Surya Siddhanta, it is stated that the star Revati is identified by all authorities as the star ζ Piscium, "of which longitude at present (A.D. 1858) as reckoned by us from the Vernal Equinox, is 17° 54′. Making due allowance for the precession, we find that it coincided in position with the Vernal Equinox, not far from the middle of the 6th century, or A.D. 570."

This estimate of the date was evidently calculated with a regression in longitude of the Equinoctial point at a mean annual rate of 50″, and the date found is no doubt a close approximation, and may be accepted as sufficient in the absence of a knowledge of the secular variation, which might produce a difference of some years, when reckoning backwards through so many centuries.

(4) We have also another calculation by Colebrooke for the age in which Brahmagupta lived, deduced from the position of Spica Virginis as given by that Indian astronomer, namely, 103° apparent longitude, from which was deduced 182° 45′ Right Ascension at his time. In 1800 A.D. the actual Right Ascension was 198° 40′ 2″ (Zach's tables), the difference of the two positions of the Equinox in Right Ascension being 150° 55′ 2″, indicated a lapse of time of 1216 years, the mean annual precession in Right Ascension being reckoned at 47′ 14″. This elapsed time subtracted from 1800, gives us the date of Brahmagupta's observation as 584 A.D.

Plate VIII.

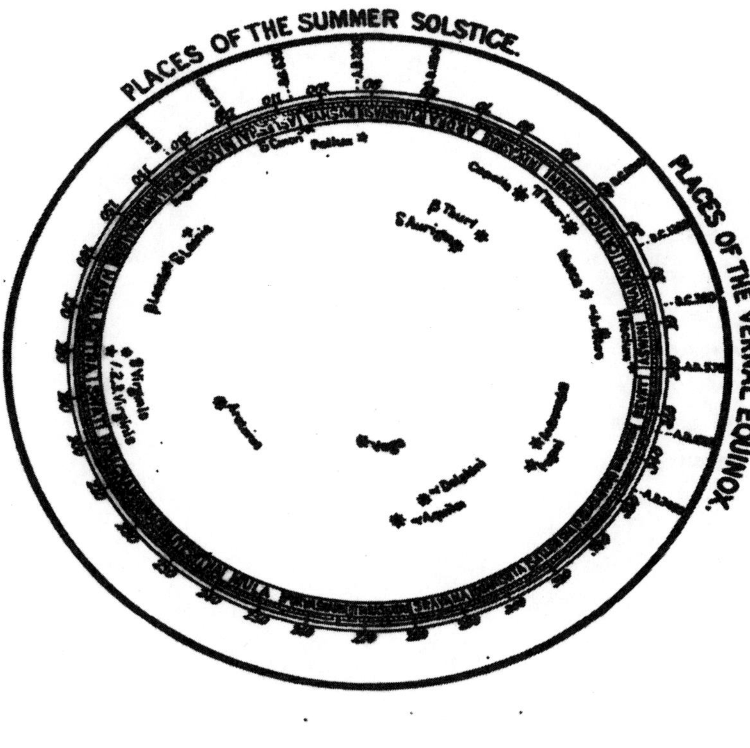

The mean of the two calculations by Colebrooke places it between 581 and 582 A.D., which is a sufficient approximation to the time when the Equinox coincided with the beginning of Aswini, and as a verification of Brahmagupta's statement relating to the position of Spica Virginis, for it may have happened at some period of his life.

It may, however, be mentioned that, among other estimates of the date when the origin of the Indian longitudes was made coincident with the beginning of Aswini, is that of Bailly, who, in his examination of the tables brought from India, deduced the time as 499 A.D., and in this opinion Sir W. Jones would appear to have coincided.

On the other hand, the Indian astronomers of Ujjaini, who were consulted by Dr. Hunter on the subject, gave 550 Saca or 628-9 A.D. as the date of Brahmagputa, which is not improbable, if he lived 45 years after the date assigned by Colebrooke.

The difference in these various estimates is to be accounted for by the very slow motion of the Equinoctial point, which is almost inappreciable in the lifetime of a man, without a very careful measurement, being less than 1° 4′ in a century.

It may, therefore, be safely assumed that the epoch when the origin of the Indian longitudes was made to coincide with the beginning of the Lunar Asterism Aswini, and of Mesha, the first of the 12 Indian signs of the Zodiac, was about 570 A.D.

To enable the reader not only to more fully understand this important feature in the Hindu system, but also to follow the argument which nearly all the Sanscrit scholars at the end of the last century employed, in tracing and establishing the authenticity of dates, as determined by the position of the Colures, the accompanying Plate VIII. is annexed. On it have been placed partial radial lines from the Pole of the Ecliptic to the points of division, which are at the commencement of several of the Nacshatras, Crittica, Bharani, Aswini, etc., showing the direction of the Equinoctial

E 2

Colure when each of these points, in succession, became the Equinoctial point, and the dates of the same, on the supposition that the mean rate of precession in longitude was 50″ (neglecting the small secular variation).

The corresponding partial lines through the Solstices, which are at right angles to these radii, have also been inserted, showing corresponding dates. The motion of this radial line is so slow that it occupies 960 years to pass from point to point through each of the 27 divisions of the Ecliptic, and when this line is referred to in the ancient Hindu writings, either as at the Equinoctial point or at a given distance from it, such reference affords evidence of the approximate date of such writings, as well as of the observation alluded to in them.

It was the opinion of nearly all who have studied the subject (Sir W. Jones, Colebrooke, Davis, Bentley, and others) that the time when the Equinoctial point was in Crittica, and for some centuries afterwards, was a period marked by considerable activity and progress in the cultivation of Hindu astronomy. Our real information regarding such astronomy, resting upon unquestionable evidence, does not, however, go much earlier than this date. What, then, was such date, *i.e.*, when the first point in Crittica was coincident with the Equinox?

Colebrooke, in his researches concerning the principal stars which give names to the Nacshatras, when describing the Asterism Crittica, refers to it as being, "now the third, but formerly the first Nacshatra"; that it consists of six stars, the principal one, according to the Surya Siddhanta, having a longitude of 37° 30′, or, according to Siddhanta Siromani and Grahalaghava, 37° 28′ to 38°, and he considered the bright star in the Pleiades, which has a longitude of 40° from ζ Piscium to be the principal star of the Asterism.*

* The Critticas are said to be six nymphs, who nursed Scanda the God of War, and he was named from these, his foster mothers, Carticeya.

When the Vernal Equinox was in the first point of Crittica, whose longitude from the beginning of Aswini (the subsequent origin of Indian longitudes), is the same as the space of two Asterisms, or 26° 40', the Southern Solstice was then at a point of 3° 20' in Dhanishtha (see Plates V., VI. and VIII.). Colebrooke, referring to Dhanishtha, says that to determine the position of this Nacshatra is important, as the Solstitial Colure, according to the ancient astronomers, passed through the extremity of it, and through the middle of Aslesha.

Bentley, so sceptical about the authors to whom the Hindus attribute their Siddhantas, and who controverted, with much industry, the opinions of Colebrooke regarding the dates of Parasara, Aryabhatta, Varah-Mihira, Bhattopala, and other Hindu astronomers, is entirely in agreement with Colebrooke as to the importance of ascertaining the period when the Southern Solstice was in Dhanishtha. He finds evidence regarding this date in the fact that the Equinoctial Colure then passed through the middle point of Visakha, thus bisecting it, from which circumstance he says it derives its name. This Colure, through the middle of Visakha, would necessarily also pass through the first point of Crittica, which would then be the place of the Vernal Equinox.

To find the date when this occurred, he calculates the difference of precession, by reference to Cor-Leonis, whose longitude from the first of Crittica is, according to all Hindu astronomers, 102° 20' (*i.e.*, reckoned from the first of Aswini). Comparing this with the longitude of the same star from the Equinox of 1750 A.D., as given in the British Catalogue of that date, viz., 146° 21', he thus finds that the Equinoxes had fallen back in the interval through an arc of 44° 1'.

Then, to find the number of years corresponding with this regression, he calculates that the mean rate of regression was 1° 23' 6.4" for each century, allowing the secular variation to be 2.27" (by which the precession is supposed to diminish each century), and the time for the whole regression through an arc of 44° 1′ would be 3176 years. If from this the date of the Catalogue be subtracted, there remains, according to such calculation, 1426 B.C., for the date when the Vernal Equinox coincided with the first point of Crittica.

This date differs by only 76 years from that which is obtained by supposing the mean rate of regression to be 50″, from the first point of Crittica to the first point of Aswini (through an arc of 26° 40′), and supposing the Equinox to have coincided with the first of Aswini in 570 A.D. We arrive, then, at this approximate date of 1426 B.C. (or 1350 B.C., according to the most recent estimate of the rate of regression) as one of the earliest which can be ascertained of any authentic facts relating to the Indian Ecliptic, a date in reference to which the earlier Hindu astronomers made their calculations.

In confirmation of the date given by him, Bentley urges that the ancient astronomers feigned the birth of four of the planets, from the union of the daughter of Daksha and the moon; the observations are supposed to be occultations by the moon, which occurred nearly at the same time in the Lunar Mansions, from which, as mothers, the planets received their names. Thus Mercury, Venus, Mars, and Jupiter were respectively called Rohineya, Maghabhu, Ashadhabhava, and Purvaphalgunibhava, the father, Soma (or the moon), being present at the birth of each, the times of their occultations being: for Mercury, 17th April, 1424 B.C.; Jupiter, 23rd April, 1424 B.C.; Mars, 19th August, 1424 B.C., and Venus, 19th August, 1425 B.C.; all within the space of 16 months.

Saturn, not then discovered, was feigned to be born afterwards, from the shadow of the earth in an eclipse of the sun, and hence called Chyasuta, the Offspring of the Shadow.

Bentley was also of opinion that the Lunar Asterisms were formed about the time when the Equinoctial point coincided with the first point of Crittica, or between the years 1528 B.C. and 1371 B.C. He finds evidence for his opinions from a statement in the Vedas, which is mentioned also in other books, where we are informed that in the first part of the Trita Yuga (this being divided into four parts), the daughters of Daksha were born, and that of these he gave twenty-seven to the moon; or, laying aside all allegory, the twenty-seven Asterisms were formed in the first part of the Trita Yuga.

One of the first steps in this astronomical method of verifying dates, by references to the regression of the Solstice, or Equinox, was made by Mr. S. Davis, who communicated to Sir William Jones a passage from the Varahi Sanhita, of which the following is, on the authority of Sir William, a scrupulously literal translation:—

"Certainly the Southern Solstice was once in the middle of Aslesha, the Northern in the first degree of Dhanishtha, by what is recorded in former Sastras. At present one Solstice is in the first degree of Carcata, and the other in the first of Macara; that which is recorded not appearing, a change must have happened, and the proof arises from ocular demonstration: that is, by observing the remote object and its marks, at the rising or setting of the sun, or by the marks in a large graduated circle of the shadow's ingress and egress. The sun, by turning back without having reached Macara, destroys the South and the West; by turning back without having reached Carcata, the North and East. By returning when he has just passed the summer Solstitial point, he makes wealth secure and grain abundant, since he moves thus according to Nature; but the sun, by moving unnaturally, excites terror."

In this passage, Varaha is explaining to some one that a change must have taken place in the position of the Solstitial Colure, which in his time passed through the two points of the Ecliptic, the first degree of Carcata, or the Hindu sign of the Crab, and the first degree of Macara, or the Sea Monster; but that in the age of a certain Muni, or ancient philosopher, the Solstitial Colure passed through the beginning of the Asterism Dhanishtha and the middle of the Asterism Aslesha. He is aware of the change, but he does not give an opinion regarding the rate of the motion, which, by observation, would have been inappreciable in the lifetime of a man (being only little more than (1¼ degrees in 100 years). He gives, however, as a fact, the position of this Colure, which was observed in his time.

The reader will easily understand the argument which Varaha employs in his proof, by noticing the points of his own horizon at which the sun rises and sets at different times of the year. On the 21st March it rises in the East point, and sets in the West, being then in the first point of Aries, the beginning of its path through the modern Ecliptic. It is seen afterwards to rise and set at points more and more towards the North, as may be observed by noticing it rise and set, behind marks, such as a tree, the spire of a church, or other objects, until, on the 21st June, it has reached a point of the horizon at which in rising it seems to stop its Northward course, and afterwards to go back towards the East point. In the interval of this Northern progress at rising, the sun has advanced in its path through three signs of the Ecliptic, and is now at the Solstice, at its greatest Northern declination, and the beginning of the modern sign of Cancer, when its Northern motion ceases. Then a Southern motion begins. There is, in fact, an apparent oscillation from North to South, within the zone of the Solar Zodiac, of the rising and setting points, or, in Hindu phraseology, "when the sun has reached a certain point of its course, it begins to turn back from the North." The apprehension of danger from the turning back Southward before the sun has reached the calculated point of his path, is, of course, a figment of the astrologers. Sir W. Jones explains that he "may have adopted it solely as a religious tenet on the authority of Garga, a priest of eminent sanctity, who expresses the same wild notion in a couplet, of which the following is a translation:—

"'When the sun returns, not having reached Dhanishtha in the Northern Solstice, or not having reached Aslesha in the Southern, then let a man feel great apprehension of danger.'"

It may be inferred from the use which Varaha makes of the Lunar

Mansions when describing the position of the Colure in the time of the Ancient Muni referred to, and the use he makes of the Solar division of signs, when explaining the position of the Colure of his own time, that the Lunar division of the Ecliptic was not in such general use as the Solar division in the time of Varaha. On the other hand, the Lunar divisions are repeatedly made use of by their names in the ancient code, the Institutes of Menu, in which only one of the signs appears to have been mentioned (where it is said that "the Sun in the sign of Kanya or the Virgin must be shunned.")

CHAPTER IV.

HINDU MONTHS AND SEASONS.

The principal method of measuring time, employed by ancient nations, was by stated revolutions of the sun, the moon, and the seasons. The apparent diurnal motions of the sun, the moon, and the stars were obvious to all mankind, and a primitive discovery, no doubt, was that all the fixed stars have one and the same uniform period in their apparent diurnal motion; but to measure the absolute lengths of each apparent period was a problem not easy of solution, and a still more difficult problem was that of reconciling unequal days, with months of unequal length, whose periods were reckoned from one new moon to the next, or from one conjunction with the sun to the next.

The difficulties experienced by the Hindus in adjusting their calendar, in which errors were so liable to spring up and increase, occasioned repeated changes of their system. At one period the motion of the moon was taken as its foundation, and the lunar month was formed to agree with the phases of the moon. Then a change took place, and a solar month was formed, constituted so as to be reckoned by the time the sun, in its progress, remained in each sign of the Solar Zodiac. Another change followed, efforts being made to reconcile the two previous systems, in which each kind of month preserved its original character, the solar month being reckoned in ordinary civil days, and the lunar months measured by tithis or lunar days, each being one-thirtieth part of a synodic period, the time elapsing between two conjunctions of the sun and the moon. The result of these efforts was the formation of the luni-solar year, reckoned either in civil days or in tithis.

From the statement of Colebrooke,* it would appear that to each Veda was annexed a treatise having the title of "Jyotish," an

---

* *Essays, Vol. I., page* 106.

astronomical work, which explains the adjustment of the calendar for the purpose of fixing the proper periods prescribed for the performance of religious duties.

In the treatises which he examined, a cycle, (Yuga) "of five years only was employed. The month is lunar, but at the end and in the middle of the quinquennial period an intercalation is admitted by doubling one month. Accordingly, the cycle comprises three common lunar years, and two which contain thirteen lunations each. The year is divided into six seasons, and each month into half months. A complete lunation is measured by 30 lunar days, some of which, of course, must in alternate months be sunk, to make the dates agree with the Nychthemera, for which purpose the sixty-second day appears to be deducted, and thus the cycle of five years consists of 1860 lunar days or 1830 Nychthemera, subject to further correction. The Zodiac is divided into 27 Asterisms, or signs, the first of which, both in the Jyotish and in the Vedas, is Crittica or the Pleiades.*

"The measure of a day by 30 hours, and that of an hour by 60 minutes are explained."

The rule upon which the method of intercalating a month, here implied, will be understood from a corresponding rule of the Siddhanta Siromani, according to which it may be deduced that in 33.53551 lunar months there are 32.53413 solar months.

To make the latter months lunar, a month will have to be added after 32 solar months, or after 2 years 8 months, and again, two months added after 5 years and four months.

From this it is obvious that a cycle of five years was too short for making the intercalation, a very much longer cyclic period being required; so that an exact number of lunar months shall

---

* It has been already explained that the date when the Equinox was at the first point of Crittica, was about 14 centuries before the beginning of the Christian era.

coincide with an exact number of solar months, and so that only a small fraction of a year or no fraction at all shall remain.

The rule in the Vedas for subtracting the sixty-second day is not quite so correct as that of Bhascara, who says that the subtractive day occurs in $64\frac{1}{11}$ lunar days (tithis).

It is a characteristic of the Hindu astronomy, distinguishing it from that of Ptolemy, that its rules are expressed rather in an analytical form than synthetical, the problems and theorems of geometry being put mostly in algebraical or arithmetical language. What we call the Pythagorean theorem assumes a variety of algebraical forms, giving solutions in integers; but the subject of Hindu algebra will be hereafter described at somewhat greater length, and reasons given for believing it to be autochthonous, and that the Arabs from whom we received it obtained it from the Hindus.

Colebrooke further makes the following important remarks:—"This ancient Hindu calendar, corresponding in its divisions of time, and in the assigned origin of the Ecliptic, is evidently the foundation of that which, after successive corrections, is now received by the Hindus throughout India.

"The progress of these corrections may be traced from the cycle of five to one of 60 lunar years, which is noticed in many popular treatises on the calendar, and in the commentary of the Jyotish, and thence to one of 60 years of Jupiter, and finally to the greater astronomical period of 12,000 years of the Gods, and a hundred years of Brahma."*

The arrangement of the 12 Hindu months, as they now stand (see Plate IX.) has, at different times been made the subject of diligent enquiry.

* It has been before remarked that the cycle of 60 years was the sosos of the Chaldeans, in use by them, according to Berosus the Chaldean Hiérarch, of the time of Alexander, from the most ancient times. In China the annals of their monarchs have been recorded in cycles of 60 years, going back as far as 2300 B.C.

Bentley, in his "Hindu Astronomy," states that the months were formed about the year 1181 B.C., when the sun and moon were in conjunction at the Winter Solstice, and that, with reference to this epoch, the Hindu astronomers had then made many improvements in their system, and, among other discoveries, they found that the Colures had fallen back through an arc of 3° 20′ from their former position in 1425 B.C., when the Winter Solstice was in 3° 20′ of Dhanishtha, but now (1181 B.C.) it was at the beginning of that constellation, and the Vernal Equinox was at a point 10° of the Constellation Bharani.

He futher states that the Lunar Asterisms which began with a month were called wives of the sun, although they had been all before allegorically married to the moon. The commencement of the year with the month Aswina, according to Bentley, was, of all others, the most celebrated.

"Durga,* the year personified in a female form, and Goddess of Nature, was then feigned to spring into existence. In the year 1181 B.C. the first of Aswini coincided with the ninth day of the moon; and on that day her festival was celebrated with the utmost pomp and grandeur. In the year 945 B.C. some further observations were made, by which they determined that in 247 years and one month the Solstice fell back 3° 20′ in respect of the fixed stars. In consequence of these observations, they threw back the epoch of the commencement of the year, with Aswina, in 1181, to the year 1192 B.C., in which year the commencement of Aswina fell on the sixth day of the moon; and the Doorga festival was ever after made to commence with the sixth lunar day of Aswina."

As a confirmation of the date of this epoch, Bentley, by a rather complicated method, calculated that the 12 months of the year were then formed and named. They were established from the

* See Plate IXa, which is a photographic representation of the marriage of this goddess, taken from an exhibition of mythological figures, held at Decca about the year 1869.

tropical revolutions of the sun, and feigned by the Hindu poets to have been born as the offsprings of the union of the moon and the Lunar Asterisms, in which the moon was supposed to be full at the time.

The name of the solar month was derived from the Asterism in which the birth occurred, and this being at the full moon, the sun would be in the Asterism diametrically opposite. Thus:—(See Plate IX.)

Month.	L. Asterism.	Month.	L. Asterism.
Mâgha	Magha.	Sravana ..	Sravana.
Phalguna ..	Uttara-Phalguni.	Bhadra ..	Purva-Bhadrapada.
Chaitra	Chitra.	Aswina ..	Aswini.
Vaisakha ..	Visakha.	Kartika ..	Critika.
Jyaishtha ..	Jyeshtha.	Margasirsha	Mrigasiras.
Ashara	Purva-Ashadha.	Pausha ..	Pushya.

The adjustment of the months to the Nacshatras may have been as here stated by Bentley, but the names of them must have been in use long before, for the months are frequently mentioned in the Institutes of Menu, in connection with the times when ceremonies, or prescribed duties, should be performed. A few examples are here extracted from that remarkable code, which every learned Brahmin is enjoined to study with extreme care, and to explain fully to his disciples, but which no man of an inferior class was permitted to teach.

"On the days of the conjunction and opposition let him (the father of a family) constantly make those oblations which are hallowed by the Gay-atri, and those which avert misfortune; but on the eighth and ninth lunar days of the three dark fortnights at the end of *Agrahayan*, (Margasirsha) let him always do reverence to the Manes of Ancestors.

"In the Month of *Aswina* let him (the father of a family of the third or fourth orders) cast away the food of sages, which he before had laid up, and his vesture, then become old, and his herbs and roots, the sun in the sign of Canya (the Virgin) must be shunned.

Plate IX

ARCS OF THE ECLIPTIC CORRESPONDING WITH
SOLAR MONTHS.

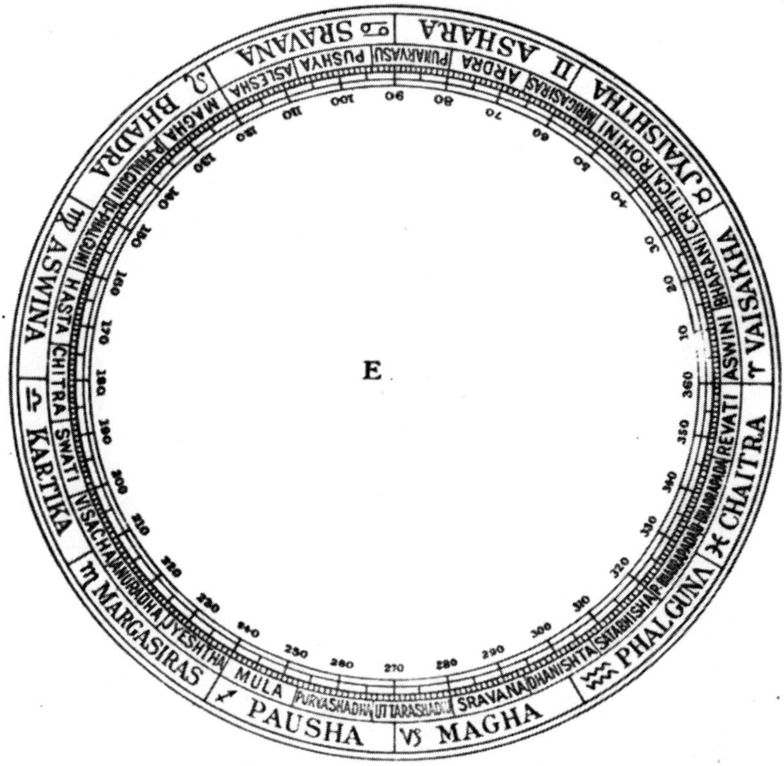

PLATE IXA.

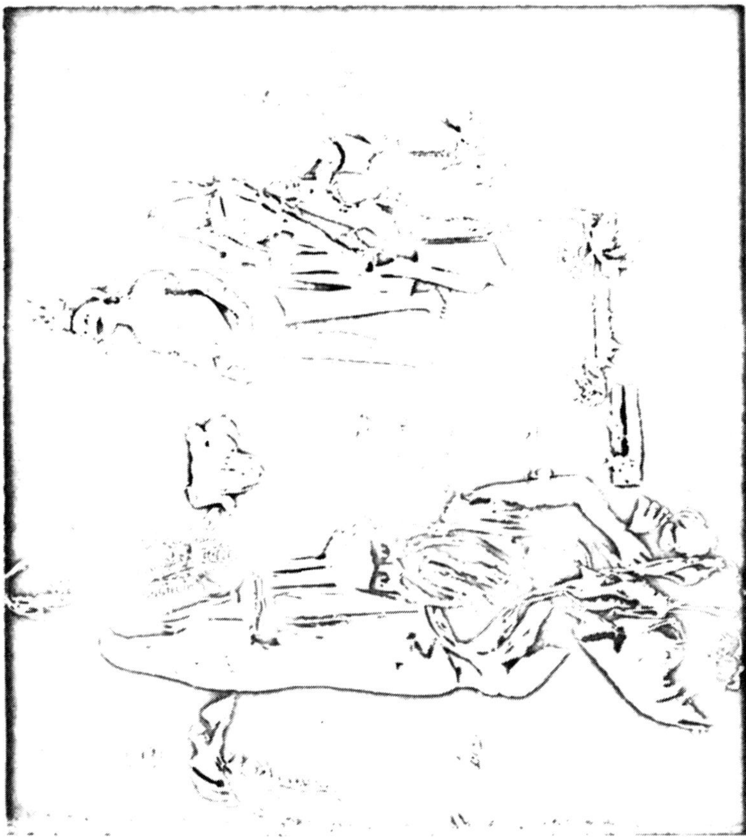

Siva receiving his consort from the hands of Kama Deva.

Photographed from one of the scenes of Hindu Mythology which were exhibited in a public spectacle in Dacca about the year 1869.

The fable is supposed to have an allegorical meaning, Siva is a personification of time, and Durga is one of many representations of the Ecliptic.

This union was considered necessary for the welfare of the universe.

"Having duly performed the Upcarma (or domestic ceremony with sacred fire) at the full moon of *Sravana* or of *Bhadra*, let the Brahmin fully exert his intellectual powers, and read Vedas during four months and one fortnight.

"Under the Lunar Asterism of *Pushya*, or on the first day of the bright half of *Maga*, and in the first part of the day, let him perform out of the town the ceremony Utserga of the Vedas.

"At the close of the season let him perform the rite called Adhvara; at the Solstices let him sacrifice cattle; at the end of the year, let his oblations be made with the juice of the moon plant.

"Not having offered grain for the harvest, nor cattle at the time of the Solstices, let no Brahmin who keeps hallowed fire and wishes for long life taste rice or flesh.

"Important duties are to be performed on some fortunate day of the moon at a lucky hour, and under the influence of a star with good qualities.

"The Shraddha, the act of due honour to departed souls, on the dark day of the moon, is famed by the name of Pitrya or Ancestral.

"The most approved lunar days for sacred obsequies are the tenth and so forth, except the fourteenth, in the dark half of the moon.

"On even lunar days, he who does honour to the manes, and under even lunar stations (*i.e.*, constellations), enjoys all his desires. On odd lunar days and under odd Lunar Asterisms, he procures an illustrious race.

"As the dark half of the moon surpasses the bright half for the celebration of obsequies, so the bright half of the day surpasses for the same purpose the dark half of it.

"The dark lunar day destroys the spiritual teacher, the fourteenth destroys the learner, the eighth and the day of the full moon destroy all remembrance of Scripture, for which reason he must avoid reading on those days.

"On the dark night of the moon, and on the eighth on the night of the full moon, and on the fourteenth, let a Brahmin who keeps house be continually chaste as a student in theology.

"When a king begins his march against the domains of his foe, let him advance gradually in the following manner against the hostile metropolis:—

"Let him set out on his expedition in the fair month of *Margasirsha*, or about the month of *Phalguna* and *Chitra*, according to the number of his forces, that he may find autumnal or vernal crops in the country invaded by him.

"On his march, let him form his troops either like a staff, or in even column like a wain, or in a wedge with the apex foremost; like a boar, or in a rhomb with the van and rear narrow and the centre broad and like a marcara or sea monster."

From these extracts it may be inferred that before the time when that ancient compendium (the Institutes of Menu) was composed or compiled, the names of the months were widely known, that they were then connected with the Lunar Asterisms, and with the established lunar synodic period, and that a long unmeasured anterior period must have elapsed before such a system could have become so universally known and established.

In the Institutes of Menu the 27 Lunar Asterisms are called the daughters of Dacsha, and the consorts of Soma, or the moon. It may also be inferred from the extract, "the sun in the sign of Canya (*i.e.*, the Virgin) is to be shunned," that the Solar Zodiac of 12 signs, though known, was not sanctioned, and all references are made to the Lunar Asterisms.

It is probable that the astronomers of the Orthodox Brahmins, who had brought with them the system of 28, or the improved system of 27, Lunar Mansions, were a sect of astronomers, separate from those who adopted the Solar Zodiac as the foundation on which their astronomy was constructed, and there may have been jealousy between the leaders of the sects, causing the Solar Zodiac to be shunned by one of them. It was also probably owing to this division amongst them that the monarchs by whom they were respectively patronised were distinguished as the Solar and Lunar races, who ruled India contemporaneously in the early ages.*

In a description of the Hindu lunar year given by Sir W. Jones in Volume IV. of his works, he says:—"The lunar year of 360 days (*i.e.*, 12 months, each of 30 days) is apparently more ancient than the solar, and began, as we may infer from a verse in the Matsya, with the month of Aswina, so called because the month was at the full when that name was imposed in the first Lunar Station of the Hindu Ecliptic, the origin of which, being diametrically

---

* The difficulties experienced by Hindu Astronomers in the division of time and in the formation of their calendar, caused principally through the erratic motions of the moon, by which all nations must at first have measured their time, a method which still subsists among Mahomedan nations and among the Chinese, were equally felt among European nations. For instance in the time of Julius Cæsar, it was ordained among the Romans, that the month should be reckoned from the course of the sun and not of the moon. The ancient solar year had consisted of 12 months each of 30 days. Altogether, 360 days; an addition of five days had been made, making it 365 days, but the tropical year exceeded this by nearly a quarter of a day, and a day was intercalated every fourth year, making that year to consist of 366 days; the year thus corrected, called the Julian year, was found in the lapse of time to be a little in excess; in the course of about 130 years, it amounted to a whole day, and in the course of about 24,000 years the seasons would be so changed that the calendar would represent Midsummer to happen in December. At the Council of Nice, held in 325 A.D., the Vernal Equinox was fixed to happen on the 21st March; in the time of Julius Cæsar it had been observed to take place on the 25th March. In 1582 A.D., the error in the calendar amounted to 10 days, and the Vernal Equinox was found to happen on the 11th March instead of the 21st March. In that year a change was effected; the 10 days in excess were taken from the month of October, the 5th day being called the 15th, thus bringing back the Vernal Equinox to the 21st March.

To prevent the recurrence of a similar error in the future, Pope Gregory XIII. effected a change and it was ordained that in Catholic countries the day beginning the century should be an ordinary year of 365 days, but that beginning the fourth century should be a leap year. But

F

opposite the bright star Chitra (*i.e.*, Spica), may be ascertained in our sphere with exactness."

Sir W. Jones also says that there is evidence of a still earlier arrangement of the months when the year was made to begin with the month Pausha, near the Winter Solstice, whence the month Margasirsha has the name Agrahayana, or "the year is next before."

"The twelve months now denominated from as many stations of the moon seem to have been formerly peculiar to the lunar year; for the old solar months, beginning with Chaitra, have the following very different names in a curious text of the Veda on the order of the six seasons :—Madhee, Madhava, Sucra, Suchi, Nabhas, Nabhasya, Isa, Urja, Sahas, Sahasya, Tapas, Tapasya."

The lunar month has a different beginning in different parts of India. In Bengal it begins at the full moon or Purnima—the midnight of the Pitris or Ancestors, who reside on the under part of the moon. By a kind of analogy, as the day of the Pitris was divided into two parts, by their midday and midnight, when the moon was in opposition or conjunction, so the month was divided into two parts, the bright half and the dark half, or, as they are called, the Sukla Paksha and Krishna Paksha, each part or paksha consisting of 15 lunar days or tithis. The tithi is defined in the

---

this correction, adopted by a statute 24, Geo. II., c. 23, in 1752, called the new style, Dr. Playfair, observes, is not the most correct, for the reformers made use of the Copernican year of 365 days 5 hours 49 minutes 20 seconds; instead therefore of inserting 97 days in 400 years they ought to have added 41 days in 169 years, or 90 days in 371 years, or 131 days days in 540 years, &c.

More recent observations have determined the tropical year to be 365 days 5 hours 48 minutes 45 seconds 30‴, and the intercalations ought to be—

Years	..	4	17	33	128	545	673	801	929	1057
Days	..	1	4	8	31	132	163	199	225	256, &c.

That is 4 days in 17 years, 8 days in 33 years, 31 in 128 years, 132 in 545 years, &c.

Surya Siddhantas as the time taken by the moon in describing 12° of the space constituting its separation from the sun. It is, therefore, $\frac{1}{30}$ of the synodic period or lunar month (the time taken for a separation of 360° from the sun).

The phases of the moon are called Calas, and they are compared to the string of a necklace or chaplet, round which are placed gems or moveable flowers; the Maha-cala is the day of nearest approach to the sun, the day of the conjunction. In the almanack it goes by the name of Amavasya, on which obsequies are performed to the manes of the Pitris, or Ancestors, to whom the darker fortnight is peculiarly sacred.*

According to the Purans (old Scriptures) the names of the months are derived from 12 of the Lunar Asterisms. The Puranics suppose a celestial nymph to preside over each of the 27 constellations, and they feign that 12 of these were consorts of the God Soma, or the moon, and that he became the father of 12 genii, or months, which were named after their respective mothers.

The mathematical astronomers, or Jyautishicas, however, maintain that their lunar year was arranged by former astronomers, the moon being at the full in each month on the very day when the sun entered the Nacshatra from which that month is denominated.

---

* An interesting passage from Quintus Curtius who, according to Niebuhr, lived under Septimus Severus, appears to show that some correct knowledge of the Hindus was known to the Romans of that time. Septimus Severus died at York in 211 A.D. It informs us that "the Indian month consists of 15 days, they indeed compute their time by the course of the moon, but not, as most other nations do, when that planet hath completed her period, but when she begins to contract her sphere into horns, and therefore they must necessarily have shorter months, who regulate their time according to this measure of lunar calculation."—*Quinti Curtii, lib.* 8, *cap.* 9.

Mr. Wilkins, in a note to his translation of the Hitipadesa, says:—"The Hindus divide the Lunar month into what they denominate the Sukla-Paksha and the Krishna-Paksha, that is the light side and the dark side of the moon; the former commencing with the new moon and the latter with the full."

F 2

According to the astronomer Nrisinha, the solar months were originally lunar, their names being derived from the Nacshatras in which the moon, departing from a particular point when they were named, was observed to be at the full; although the full moon did not always happen in those particular Nacshatras, yet the deviation never exceeded the preceding or succeeding Nacshatra, and whether it fell in Hasta, Chitra, or Swati, still that month was named Chaitra, and so of the rest.

THE SEASONS.

The Hindu solar year is divided into six seasons, each consisting of periods of two months, or whilst the sun remains in two signs successively.

The very cold season named Sisira, is reckoned from the time when the sun is in the Winter Solstice, it is followed in order by:

The Spring named Vasanta.

The Hot Season named Grishma.

The Rainy Season named Varsha.

The Autumn Season named Sarat.

The Cold Season named Hemanta.

CHAPTER V.

THE RISHIS.

The reader being now in some measure acquainted with the nature of the Indian Ecliptic, will be able to form a conception of the accuracy with which observations on the Colures could be made by taking as an example one of the most simple and ancient methods employed for ascertaining the day on which the sun was in the Summer Solstice.

The bright star Regulus, whose longitude is now about 148° from the Vernal Equinox, is the principal star of the Indian Lunar Asterism Magha; it was close to the Summer Solstice in the year 2280 B.C., being only 27′ north of it. Consequently, the sun, in its annual course through the Ecliptic, would be in the Solstice when passing the star Regulus, it being then only ¼ of a degree from the sun's upper limb. Observations of the heliacal rising of Regulus (shortly before or after) would give fairly accurate results of the place of the Solstice, especially if the observations were carried on for many years, and the retrograde motion of the Solstice must necessarily have been discovered about that time, if, indeed, it had not been discovered long before. The importance always attached to the sacred days when the sun was in a Solstice has been before referred to as intimately associated with religious ceremonies, and allusions to a time when the Summer Solstice was in the Constellation Leo are conspicuous in the ancient writings found in most Eastern countries. It is, however, certain that the fixed star Regulus, the principal one of that constellation, marked the position of the Summer Solstice in or about the year 2300 B.C., and then to all nations, the sun would be seen to rise together with that star. For the period of 200 years, both before and after that date, the Solstitial point, in its slow motion along the Ecliptic, would not be more distant than 3° from the star, and during these 400 years, therefore, it would be seen to rise at midsummer shortly before or after the sun. Hence, Regulus, under the name of Magha in India, must have been a star of considerable importance, not only to the Indo-Aryans, before they arrived in India, but also to all the Asiatic tribes, for, being a fixed point close to the Ecliptic, it was most convenient to all, as pointing the sacred days of the year, and as affording estimates of the longitudes of the moon and planets. When a planet was in conjunction with that star, from this circumstance alone it was known that it had a "longitude of 9° in Magha." The Solstice in its retrograde movement over the 9° would have been at the beginning of the Constellation Magha, at or about the year 1590 B.C.

Now, at this period, the ancient Hindu astronomers evidently assumed "the beginning of Magha" as a point in the great circle of the celestial sphere passing through the Pole of the Ecliptic, as a starting position for the Solstitial Colure, from which to reckon its retrogression, and they called this "the line of the Rishis." This line remained fixed whilst the Solstitial Colure continued to retrograde.

Students of the astronomical writings of the Hindus have been more or less puzzled with certain assertions found in them, relative to the alleged motion of the stars, known as "the Rishis." The Hindu commentators themselves, when treating of these supposed motions, have apparently either not understood the full meaning of the astronomers, whose doctrines they were referring to, or have enveloped their own statements in some amount of uncertainty and confusion. With a view to clear up a question which has occasioned considerable controversy, it is necessary to briefly examine the subject of the Rishis, and to offer an explanation.

Mystery, indeed, hangs over most of the ancient writings of the Hindus. Many of their Scriptures, which we believe to be the productions of living men, are ascribed by them to the Gods.

The real authors of these sacred writings were, no doubt, pious men who concealed their authorship, or with humility disclaimed the merit of their work, as being due to the Supreme Being rather than to themselves. To this assumption, their countrymen yielded their assent, and they were accordingly deemed to be inspired saints.

These were the Rishis, and Munis, men gifted above others of their race, and devoted to lives of meditation and contemplation of the Deity, and even seeking absorption in the same spiritual essence. They figure by name sometimes as Anchorites and Sunyasees whilst on earth, and afterwards as still existing, amidst the constellations as single stars in the celestial sphere.

By the word Rishi*, according to Colebrooke, is generally meant the inspired writer, or the saint of the text, the person to whom the passage was revealed, or "the author, notwithstanding the assertion of the Hindus that the Vedas were composed by no human authors."

It is a singular fact, says Sir W. Jones, that "in the Sanscrit language, Ricsha means a constellation, and also a bear, so that Maharesha may denote either a great bear or a great Asterism.

---

* From a root which meant "to shine" the Seven (Rishis) Rikshas or shiners received their name; and to the same root probably belongs the name of the Golden Bear; the Greek ἄρκτος and Latin Ursa, as the Germans gave to the Lion Goldfusz; and thus when the epithet had by some tribes been confined to the bear, the seven shiners were transformed into seven bears, then into one bear with Arcturus for their bearward. In India also, the meaning of Rikaha was forgotten; but instead of referring the word to bears, the people confounded it with Rishi, wise, and the seven stars or shiners became the abode of seven sages or poets. The same lot befel another name for this constellation. They who spoke of the seven triones had long forgotten that their fathers spoke of the stars as taras (staras) or strewers of light, and converted the bearward into Bootes, the ploughman: while the Teutonic nations, unconscious that they had retained the old root in their word stern or star, likewise embodied a false etymology in wagons and wains.—*Max Muller, Lectures on Languages, Second Series VIII., Westminster Review, January* 1865, *p.* 48.

Egyptologists may perhaps derive the Megas Arctos of the Greeks from an Indian compound ill understood."

The Megas Arctos of the Greeks is the well-known constellation Ursa Major of the Romans, the Great Wain, as it was called in every age of astronomy, to distinguish it from the lesser Wain, the constellation Ursa Minor. Diodorus Siculus informs us that travellers through the sandy plains of Arabia directed their course by the bears, in the same manner as navigators guide their vessels at sea.

Now, in the Indian astronomy, the seven stars of Ursa Major from α to η are called respectively Cratu, Pulaha, Pulastya, Atri, Angiras, Vasishtha, and Marichi, and these are the names of the seven sages known collectively as "the Rishis," so frequently mentioned in their most ancient writings.

Vasishtha, according to the Index Anucramani of the respective authors of each passage of the Vedas, was the composer of the hymns contained in the seventh book of the Rig Veda; Atri of those in the fifth; Angiras of those in another; Marichi was the father of two other composers of books of the same Veda; he was, as the name implies, the Great Rishi. Other composers were Vyasa, the son of Parasara, and several descendants of Angiras.

These seven stars, denoted by their Indian names, are shown in the positions they occupy with respect to the 27 Asterisms of the Indian Zodiac, in Plate X. They have been projected on the supposed plane of the Equator, in accordance with the method previously described, from the latitudes and longitudes given in the following extract, from the Brahma Siddhanta of Sacalya:—

"At the commencement of the Yuga, Cratu was near the star sacred to Vishnu (Sravana), at the beginning of the Asterism. Three degrees East of him was Pulaha, and Pulastya at ten degrees distant from this; Atri followed, at three degrees from the last; and Angiras at eight degrees from him; next came Vasishtha, at the distance of seven degrees; and, lastly, Marichi, at ten. Their

Plate X.

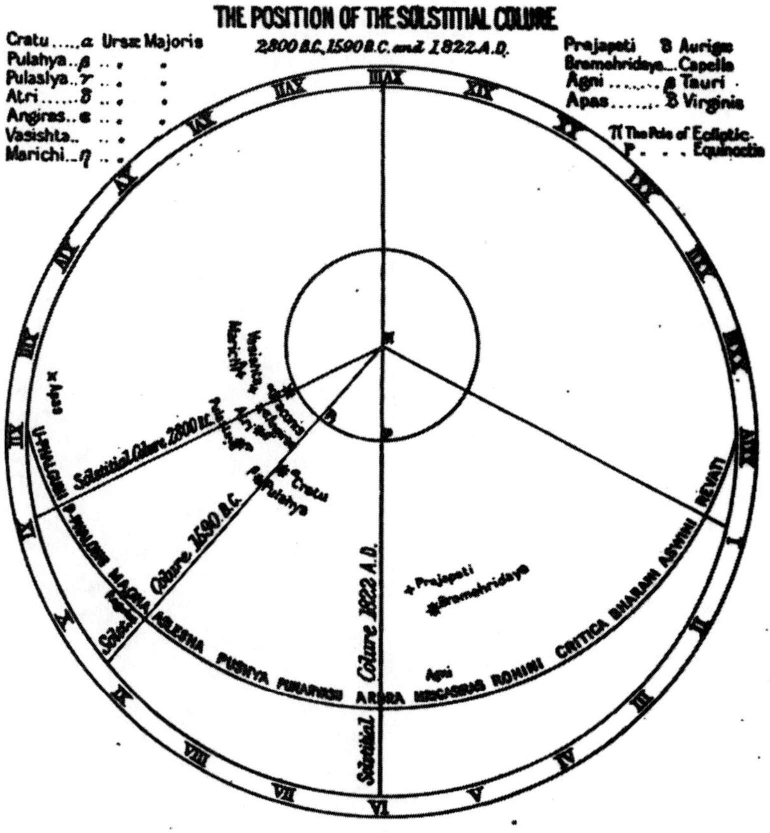
THE POSITION OF THE SOLSTITIAL COLURE
2800 B.C. 1590 B.C. and 1822 A.D.
Cratu .....α Ursæ Majoris
Pulahya..β
Pulastya..γ
Atri......δ
Angiras..ε
Vasishta..
Marichi..η
Prajapati δ Aurigæ
Bramahridaya...Capella
Agni ....β Tauri
Apas........δ Virginis
π The Pole of Ecliptic.
P . . . Equinoctia
Solstitial Colure 2800 B.C.
Colure 1590 B.C.
Colure 1822 A.D.
Solstitial
U-PHALGUNI
P-PHALGUNI
MAGHA
ASLESHA
PUSHYA
PUNARVASU
ARDRA
ROHINI
CRITICA
BHARANI
ASWINI
REVATI
Cratu
Pulahya
Prajapati
Bramahridaya
Agni
Apas

motion is eight minutes in a year. Their distances from the Ecliptic North were, respectively, 55°, 50°, 50°, 56°, 57°, 60°, and 60° For moving in the North the sages employ 2,700 years in revolving through the assemblage of the Asterisms, and hence their positions may be easily known at any particular time."

Now, the peculiar motion ascribed in the Siddhantas and other works to these seven fixed stars, the Rishis (which motion has no real foundation), has excited a considerable amount of discussion.

Extracts are given by Colebrooke from the works of no fewer than twelve different Hindu authors, all of whom were of the opinion that the Rishis had the motion referred to, and it was supposed to occupy them 100 years in their progress from East to West over the space allotted to each Asterism along the Ecliptic. The supposed motion is not noticed, however, by the Surya Siddhanta or by its commentators. Nrisinha rejects the rule of computation given for estimating the motion as not agreeing with the Puranas. Bhascara, according to Muniswara, omitted this topic, on account of contradictory opinions concerning it, and because it is of no great use.

Muniswara, in his own compilation, the Siddhanta Sarvabauma, observes that the seven Rishis, are "not like other stars, attached by spikes, to the solid ring of the Ecliptic, but revolve, in small circles, round the Northern Pole of the Ecliptic."

"Camalacara notices the opinion delivered in the Siddhanta Sarvabhauma, but observes that no such motion is perceptible, remarking, however, that the authority of the Puranas, and Sanhitas, which affirm their revolution, is incontrovertible, he reconciles faith and experience, by saying that the stars themselves are fixed; but the seven Rishis are invisible Deities, who perform the stated revolutions in the period specified."

Regarding this explanation, Colebrooke dryly remarks, if Camalcara's notion be adopted, no further difficulty remains, but it could hardly be supposed that the celebrated astronomers, Lalla and Varaha Mihra, who were not mere compilers and transcribers, intended to describe revolutions of invisible beings, and it can scarcely be supposed that they should have exhibited rules of computation, which did not approach to the truth, at the very period when they were proposed.

From the extracts above given, it will be seen that the several writers refer to a motion which they themselves evidently did not understand, but which they were endeavouring to explain from traditional doctrine, received from previous astronomers, to whom the subject was really clear.

Before, however, proceeding to the explanation of what was meant by the original Hindu astronomers, in regard to this supposed motion of the Rishis, as represented in the quotations from modern authors of Hindu astronomical works; for the benefit of general readers it may be advantageous to give some account of the theory of modern astronomy relating to the same subject.

For convenience of reference, and to make the subject more easily understood by the reader, a modern map (Plate XI.) of the Northern Hemisphere is here given. It is supposed to be a projection of the principal stars on the plane of the Equator, as that plane is now situated, the divisions on the circumference representing right ascensions, and the radii being each supposed to be divided into 90 equal parts to represent degrees.

Now, the relative positions of the stars remain constant and unchanged, however remote the time may be.

From this map Plate X. has been drawn, showing the positions of the seven stars of Ursa Major, with their Indian names, and connected with them the unchangeable position of the Nacshatras, as nearly as it can be ascertained with certainty. The Pole of the Equator P, moving in a contrary direction to the signs of the Zodiac round π, (the Pole of the Ecliptic), the great moveable circle the Solstitial Colure, passing through the two Poles π, P in the course of one revolution, must, of course, have been in coincidence with

PLATE XI.

Northern Hemisphere.

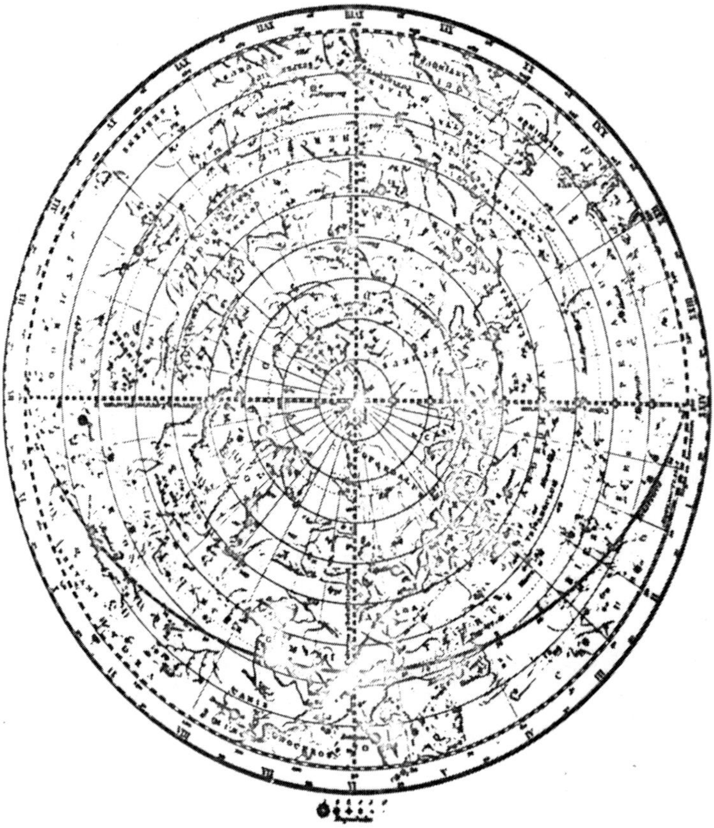

every star of the Celestial Sphere. In 4248 B.C. it was in coincidence with Marichi (or η Ursæ Majoris), whose longitude from the Equinox of 1894 was 175° 18′ 10″, then after passing over ζ in turn, it was nearly in coincidence with ε, in 3510 B.C., when the Solstitial point was at the beginning of the U-phalguni. Nine hundred and sixty years later it arrived at the beginning of Purva-phalguni, and entered the Constellation Magha (at the end of it), then, after the lapse of a further period of 960 years, it reached the beginning of that constellation.

The Solstitial Colure then coincided with the fixed unchangeable circle, assumed and called the line of the Rishis, passing through that fixed point (the first of Magha) and the Pole of the Ecliptic in the year 1590 B.C., and in about 335 years later, or in 1255 B.C., leaving the line of the Rishis in its retrogression, it coincided with Cratu or α Ursæ Majoris. This is an epoch before referred to by Davis, Colebrooke and Bentley, but reckoned by them to have been at a little earlier date (see p. 54).

The longitude of α Ursæ Majoris, deduced from the Right Ascension and Declination, as given in the Almanacs of 1894, is 133° 44′ 25″, and the longitude of the first of Magha is 138° 23′ 20″, this shows that the Colure, when it coincided with Cratu was 4° 38′ 55″ from the position which it occupied when it coincided with the beginning of Magha, *i.e.*, the line joining it with the Pole of the Ecliptic—the line which was technically called by the ancient astronomers "the line of the Rishis." When the two lines coincided, it would appear that they both went by the same name—the one fixed and the other moveable—from which circumstance the more modern Indian astronomers have confounded the lines, and, following the one that was moveable, made a supposed movement of the stars themselves.

It will be observed that a line drawn on the figure from the Pole of the Ecliptic π to the beginning of Magha, lies between α and β, which agrees with the statement of the commentator Sridhara

Swami. Next, with regard to the further statement of this writer, and of others who were of opinion that this line of the Rishis was in motion, and that it remained in each Asterism a hundred years:

It will be observed that the astronomers of the period between the 10th and 14th centuries before the Christian Era had made many discoveries, and amongst others this, that the Solstitial Colure was moving backwards along the signs. Approximate values of the rate of motion were computed, which computations resulted, as stated by Bentley, in their finding that in 945 B.C. the Solstices had fallen back 3° 20′ in respect of the fixed stars during a period of 247 years and one month, from the position they had in the year 1192 B.C. This makes the mean annual rate of motion backwards 48·56661″. Now, neglecting the decimal part of this value of the regression, which would express what was to be stated in round numbers, and reducing the arc of an Asterism, or 13° 20′, to seconds, it is seen that there are exactly 48,000″ in an Asterism, and, dividing this by the annual rate, 48″, it would take just one thousand years for the Solstitial point to travel over it. In actual fact, it takes 960 years, as previously stated.

Now, what is more natural than that omissions or mistakes should be made in the numerous copies of the statements of the original astronomers, who lived more than 28 centuries ago, or that a cipher should have been lost, or even a dot (which, we are told, ancient writers used in lieu of a cipher), at the end of the number, and that modern Hindu writers should have been misled in stating 100 instead of 1,000 years, 2,700 years for a revolution instead of 27,000? With a mean value of 50″ for the precession, we reckon 25,920 years for the revolution of a Solstice or of an Equinox.

In the preceding passages with respect to the Rishis, quoted by Colebrooke from various astronomical works of the Hindus, the writers agree in the common mistake of a supposed motion of the line of the Rishis, and in the opinion that a Solstice moves through each Asterism in 100 years; but we can only regard these mutilated fragments of a nearly perfect theory as having had a common origin, in a remote age. We may suppose that they have been handed down from the same Jyotish family, by its scattered descendants, and that the original doctrines have lost their true form, from repeated transcripts, during long periods of time, and this liability to error would be increased by the complex nature of the subject without sufficient explanation. In short, the rate of motion of the Solstices, originally known and so near to the truth, became lost to the successors of the earliest astronomers.

HINDU THEORY OF A LIBRATION OF THE EQUINOXES.

To the theory of a revolution of the Colures there was a rival doctrine, which may have been the cause of the former theory becoming neglected, and, in a great measure, forgotten. This was the doctrine of a libration of the Equinoctial and Solstitial points. Colebrooke, in his essay on the equinoxes, has given the views of a number of writers on the subject; by some the motion is considered to be an entire revolution, through the whole of the Asterisms; by others, and those the most numerous, it was a libration, between certain limits on each side of a fixed point; by a few, amongst whom was the celebrated astronomer Brahmagupta, who (though he was aware of the fact that the Southern Solstice had been formerly in the middle of Aslesha, and the Northern in the beginning of Dhanishtha) had doubts regarding the motion. He remarks upon the passage in the text relating to their former position, "this only proves a shifting of the Solstices, not numerous revolutions of it through the Ecliptic."

It will not be necessary in this connection to give more than two extracts from authorities who have assumed the doctrine of a libration. A passage from Bhascara's description of the Armillary Sphere states that:—

"The intersection of the Equinoctial and Ecliptic circles, is the Cranti-Pata, or intersecting point of the sun's path. Its revolutions, on the authority of Surya, are retrograde, three myriads in a calpa."

This is the same with the motion of the Solstice, as affirmed by Munjala and others.

The following is the corresponding passage from the Surya Siddhanta :—

"The circle of Asterisms moves Eastward 30 score in a Yuga."

In a later translation by Pundit Papu Deva Sastri, the passage is thus rendered :—

"The circle of Asterisms librates 600 times in a great Yuga," and the translator, in explanation, proceeds thus :—

"(that is to say) all the Asterisms at first move Westward 27°. Then, returning from that limit, they reach their former places; then, from those places they move Eastward the same number of degrees, and returning thence, come again to their own places. Thus, they complete one libration, or revolution, as it is called. In this way the number of revolutions in a Yuga is 600, which answers to 600,000 in a Kalpa."

Now, Bhascara was too good a mathematician to have made the mistake of putting 30,000 for half of a revolution, or for the retrograde motion of the libration, instead of 300,000. There must, therefore, have been some mistake in the transcript or in the translation.

In these two statements it may be noticed that Bhascara supposes the Equinoctial point is in motion, whereas the Surya Siddhanta assumes that the entire circle of the Asterisms oscillates, first 27° on one side of a mean point, and then 27° on the other side of that point. This supposed motion of the whole of the constellations may have led Bentley to assume that the ancient astronomers had two systems of Lunar Asterisms, the one fixed and the other moveable, the latter of which he called the Tropical Sphere, which was at one time in coincidence with the Sidereal Sphere, and from this it has been separating, at a rate equal to the annual precession.

The theory of a libration, as expressed in various astronomical works, has been shown by Celebrooke to have been generally prevalent from very early times. It was also a doctrine maintained by Aryabhatta and Parasara, and by most of the Hindu astronomers of later times.

The conception of a vibration was, without doubt, suggested by the peculiar motion of the Pole of the Equator about the Pole of the Ecliptic.

The choice of 27° was obviously an arbitrary selection for the limit of a libration on each side of a mean point. The arc of a Nacshatra of 13° 20′ would not have served the purpose so well, for connecting the motion with the Calpa. Other arcs which might have been made use of did not lend themselves conveniently to the construction. The number 27 is the same as that of the Lunar Mansions, double that number is the same as that of the seconds, in the mean annual rate of motion in the libration, namely, 54″.

If we view this subject in its connection with the geometry of the sphere, we see still further evidence of design, in the choice of 27° as the limit of the vibration.

Let the large circle E S C in the figure represent the Ecliptic in the plane of the paper. The centre π being the projection of its pole, and let A P P′ be the projection of the small circle in which the Equinoctial Pole is moving round π, in the direction indicated by the order of these letters.

Suppose P to be the place of this Pole at any time, then the great circle π P S joining the two Poles will represent the position of the Solstitial Colure at that time, and S, the point where it intersects the Ecliptic, will in this figure represent the Summer Solstice, being nearest to the Equinoctial Pole.

Now, if we assume the arc of the Ecliptic S T to be equal to 27°, *i.e.*, T to be the limit of a vibration of the Solstice from S, and suppose a great circle of the sphere T O P¹, to be drawn so as to touch the small circle in P¹, intersecting the Colure π P S in O, then O would be a point about which, as a fulcrum, any other great circle joining it with the Equinoctial Pole, would oscillate, and the points of intersection of such a circle with the Ecliptic would librate within the prescribed limits. For example, let *p* O *t* represent an

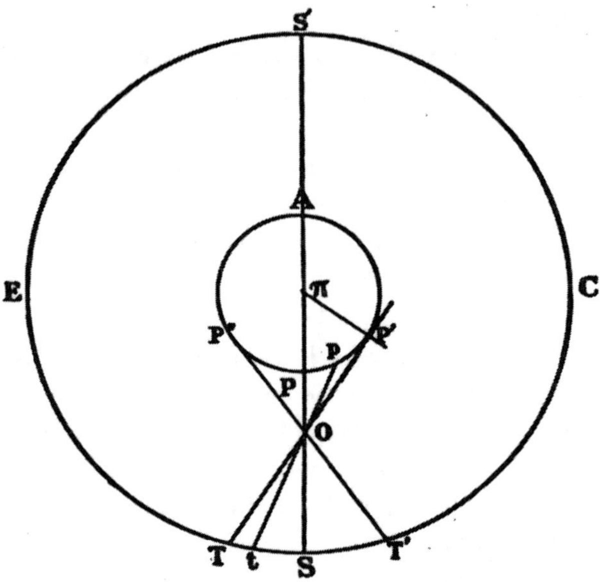

arc of such a circle, meeting the Ecliptic in *t*, then *t* would librate between the two supposed limits of 27° on each side of the mean point S, and would complete a revolution in the same time as the Equinoctial Pole would describe a revolution round π.

It will be observed that, in accordance with the hypothesis of a limit of 27°, the figure shows that there are two right-angled spherical triangles T S O and π P' O, in one of which S T is given 27°, in the other π P' is the measure of the obliquity, and, according to Hindu Astronomers, equal to 24°; also the vertical angles at O are equal. From these conditions the distance of O from the Ecliptic may be determined, and we find

$$O\,S = 45^\circ\ 33'\ 6''.$$

And the angle P π P' or the arc P P', will be 63°, or the complement of the limiting arc 27°. Thus, while the Pole moves through the arc P′ A P″, or 234°, *t* will be moving from T to *t*′ in a retrograde direction through 108°, but through the same arc in the direction of the signs, while the Pole moves from P″ to P′.

The answer to the question, "How could a libration of the Equinoxes according to the hypothesis of the Hindu Astronomers be explained and reconciled with our own theory of their motion?" was first sought in the solution of the two spherical triangles T S O and ♈ P′ O which gave the two formulæ—

$$\tan OS = \sin ST \cot \text{♈}P'. \quad (1.)$$

$$\cos P'\text{♈}O = \tan \text{♈}P' \tan OS. \quad (2.)$$

Multiplying (1) and (2) together, and suppressing common factors, we have—

$$\cos P'\text{♈}O = \sin ST$$

or the angle P′ ♈ O is the complement of the arc T S. This is a general solution which would apply to other arcs of libration besides that of 27°.

To return to the subject of the "line of the Rishis," the reader will observe that the suggestion made to the effect that the real meaning of that expression was obscured to, and by, the later astronomers in India, is strengthened by a consideration of the theories of such later astronomers, particularly of that relating to the libration of the Equinoctial point.

When it is considered that the earliest Hindu astronomers regarded the whole starry firmament as fixed, and accordingly framed their Ecliptic, as a fixed dial, peculiar to their system, it is easily seen that the Solstitial point and Colure (to them so important) was as an index finger moving on such dial. Further, when it is considered that they definitely announced, at intervals, that the Colure had retrograded so many degrees, minutes, and seconds since a previous occasion, giving the annual rate very closely to the rate now accepted as the truth, we are driven to the conclusion that the "line of the Rishis" was a part of their fixed system, and was a line represented by the Solstitial Colure as it was at the date when they fixed it, *i.e.*, when the Solstitial point was coincident with the first point of the Lunar Asterism, Magha. The date when it was so coincident was in 1590 B.C. Thus, the "line of the Rishis" remained fixed, as the "first of Magha," and thus "the line of the Rishis in Magha" was a datum from which to reckon in those ante-historic times by means of the moving Colure which separated more and more from it; just as subsequently, in A.D. 570, the "first of Aswini" was made by later astronomers a fixed datum on the Ecliptic from which to reckon their apparent longitudes. In conclusion, it may be stated that, although probably the epoch 3102 B.C. (the commencement of the Kali Yuga) was an epoch arrived at by calculation backwards; yet the epoch 1590 B.C. was one fixed by observation of the then astronomers, and always referred to subsequently by allusion to "the line of the Rishis," then established.

In a subsequent chapter, which deals with the decadence of Hindu astronomy whilst the nation was under Buddhist influence, and whilst the Hindu religion (with its astronomical and mathematical accessories) was under a cloud, an endeavour will be made to explain how, probably, the ancient accurate astronomical knowledge must have, to some extent, been lost.

CHAPTER VI.

THEORY REGARDING THE CAUSES OF THE PLANETARY MOTIONS, &c.

It is natural for men to form theories to account for the phenomena of the Universe.

Guided by appearances in regard to surrounding objects, the earth itself was thought at first to be a vast plane over which the sun, the moon, the planets, and the stars seemed each to perform a daily course; and when it was found that the true form of the surface was more nearly that of a sphere, it was still no easy matter to account for the manner by which it was upheld in space, with all the celestial bodies moving round it daily, except by reference to supernatural agencies.

For thousands of years this apparent motion of the sidereal system, or, in reality, the actual diurnal motion of the earth round its axis, has been going on uniformly, without sensible variation; although the axis itself has undergone constant change in position by a slow conical motion, requiring nearly 26,000 years to complete one revolution.

The astronomers of the Siddhantas, influenced, no doubt, by their reverence for the sacred writings and the fear of offending caste prejudices, say very little regarding the causes of the planetary motions, beyond giving a general statement of them, as understood by the more ancient astronomers.

The common opinion was that the sun and the planets, with the stars, were carried diurnally Westward by a mighty wind or æther, called Pravaha, which was moving continuously in a kind of whirling vortex.

It was supposed that the apparent Eastward motion of the planets in their orbits was brought about by an overpowering

e 2

influence of the stars, causing them to hang back, and that the irregular motions were produced by invisible Deities at the apogees and the nodes of the different orbits, those at the apogees attracting them unequally by means of reins of winds, thus guiding them in their course, whilst the others, situated at the nodes, deflected to the North or the South of the Ecliptic.

The notion that the planets were carried by an æther whirled about the sun (however ridiculous it may appear in the light of modern science) was one also prevalent in Europe before the times of Kepler and Newton. Even Descartes and Leibneitz and a crowd of followers bestowed much labour and extensive learning in proving the system of vortices to be a necessity; and it was not till long after the publication of the Principia, that the Cartesian doctrines were abandoned at Cambridge.*

Bhaskara, in his Siddhanta Siromani, after giving a number of reasons proving that an eclipse of the moon is caused by its entering the shadow of the earth, and that the sun is eclipsed by its being covered by the moon, as with a cloud, goes on to say:—

"Those learned astronomers who, being too exclusively devoted to the doctrine of the sphere, believe and maintain that Rahu cannot be the cause of the obscuration of the sun and moon, founding their assertions on the above-mentioned varieties, and differences in the parts of the body first obscured, in the place, time, causes of obscuration, etc., must be admitted to assert, what is at variance with the Sanhita, the Vedas, and the Puranas.

---

* The Principia was first published in 1687.

David Gregory gave instruction upon the Newtonian Philosophy in Edinburgh for several years prior to his removal to Oxford in 1690.

Whiston, in the memoirs of his own life, says, referring to him:— "He had already caused several of his scholars to keep acts, as we call them, upon several branches of the Newtonian Philosophy, while we at Cambridge (poor wretches) were ignominiously studying the fictitious hypotheses of the Cartesians."

The Physics of Rohault were in use to a much later period than this.

"All discrepancy, however, between the assertions of the above referred to and the sacred Scriptures may be reconciled by understanding that it is the dark Rahu which, entering the earth's shadow, and which, again entering the moon in a solar eclipse, obscures the sun by the power conferred upon it by the favour of Brahma."

It was usual with the authors of the Siddhantas to give the fabulous description of Hindu Cosmography as set forth in the Vedas and the Puranas, though they themselves might not be at the pains to assert their faith in it. Bhaskara, with great patience, goes through the account of the six Dwipas and the seven seas of milk, curds, clarified butter, sugar-cane juice, wine, and sweet water; the positions of the mountains in Jumbu Dwipa and the nine valleys, the Golden Meru, the abode of the Gods, the gardens, the lakes, and rivers in which the celestial spirits, when fatigued with their dalliance with the fair Goddesses, disport themselves.

But he himself attaches no credit to what he describes, and he concludes with the words: "What is stated here rests all on the authority of the Puranas."

He thus reasons regarding the various supporters of the earth:—

"If the earth were supported by any material substance or living creature, then that would require a second supporter, and for that second a third would be required. Here we have the absurdity of an interminable series. If the last of the series be supposed to remain firm by its own inherent power, then why may not the same power be supposed to exist in the first, that is, the earth?"

He asserts that the earth has an inherent power of attraction:—

"The earth attracts any unsupported heavy thing towards it. The thing appears to be falling, but it is in a state of being drawn to the earth. The ethereal expanse being equally outspread all around, where can the earth fall?"

To the Bauddists, who assert that the earth is going down eternally in space, he says:—

"Observing, as you do, O Baudha, that every heavy body projected into the air comes back again to, and overtakes, the earth, how, then, can you idly maintain that the earth is falling down in space (thinking that the earth, being the heavier body, would go faster and would never be overtaken by the lighter)?"

To the Jaina, who is a heretic, and disliked by the Brahmins, he says:—

"But what shall I say of thy folly, O Jaina, who, without object or use, supposest a double set of constellations, two suns, and two moons? Dost thou not see that the visible circumpolar constellations take a whole day to complete their revolutions?

"If this blessed earth were level like a plane mirror, then why is not the sun revolving above at a distance from the earth, visible to men, as well as to the Gods (according to the Puranas the sun is always revolving about Meru above the earth and horizontally)?

"If the Golden Meru is the cause of night, then why is it not visible when it intervenes between us and the sun? And Meru, being admitted by the Puranas to lie to the North, how comes it that the sun rises (for half the year) to the South?"

This and the like reasonings of the authors of the Siddhanta Siromani, exhibit a keenness of observation which would do credit to latter-day European philosophy.

CHAPTER VII.

ARITHMETIC, ALGEBRA, AND GEOMETRY OF THE HINDUS.

In histories of the mathematical sciences it has been usual to trace our knowledge of arithmetic to the Arabs, and our numerals are distinguished from those of the Greeks and Romans by the symbols termed Arabic. Dr. Peacock, in his work on arithmetic, observes there is nothing in the Greek notation which in the slightest degree resembles our own, and nothing in the object proposed in the researches of Archimedes and Apollonius which could naturally lead to its invention.

In Bhascara's Vasana, it is stated that, according to the Hindus, numeration is of divine origin, "the invention of nine figures, with the device of places, to make them suffice for all numbers, being ascribed to the beneficent Creator of the Universe."

Dr. Peacock remarks upon this passage:—"Of its great antiquity amongst them there can be no doubt, having been used at a period anterior to all existing records.

"Most other memorable inventions they have attributed to human authors, but this, in common with the invention of letters, they have ascribed to the Divinity, agreeably to the practice of the Greeks, Egyptians, and most other nations, with respect to more important inventions in the arts of life whose origin is lost in the remoteness of antiquity."

The Sanscrit names of the ten numerals are:—

1.	Eca.	6.	Shata.
2.	Dwau.	7.	Sapta.
3.	Traya.	8.	Ashta.
4.	Chatur.	9.	Nova.
5.	Ponga.	10.	Dasa.

"These have been adopted, with slight variations, not merely in all languages of the same class and origin, but likewise in many others which are radically different from them. If we proceed to the expressions of higher numbers, we find the same general law of their formation by the combination of names of the articulate numbers, with those of the nine digits.

"From consideration that when a national literature, whether oral or written, is so generally diffused as to form a standard, or a test of purity, which, while it enforces a legitimate character upon all existing terms, watches over the introduction of all others with extreme jealousy; from this consideration alone, independently of other evidence, we should be inclined to assign to the Sanscrit terms for high numbers, and, consequently, to their system of numeration upon which they are founded, an antiquity at least as great as their most ancient literary monuments; as the arbitrary impositions of so many new names for the most part independent of each other, and in numbers, also, so much greater than could possibly be required for any ordinary application of them, would be a circumstance entirely without example in any language which had already acquired a settled and generally recognised character."

ALGEBRA.

It has been usual to ascribe the origin of Algebra also to the Arabs, but there is little doubt that it is as old as any knowledge that we possess, for it is a natural method by which the mind investigates truth.

The name Algebra is supposed by some to be derived from Arabic words; by others from a supposed inventor whose name was Geber, to which the particle Al is added, making Al Geber, signifying in Arabic, the reduction of fractions to integers.

Peter Ramus, in his Algebra, says the name Algebra is Syriac, signifying the art and doctrine of an excellent man, and that there was a certain learned mathematician who sent his algebra written in the Syriac language to Alexander the Great, and he named it Almu Cubala, that is, the book of dark and mysterious things.

Indications of the science are traceable in the writings of the ancient philosophers, whose contemplation of nature required such an aid.

The earliest Arabic work on algebra, written by a Mahomedan, is, as declared by themselves, a treatise which was the production of Mahomed Ben Musa, of Kowarezm, in the reign of the Caliph Al Mamun, son of the famous Caliph Haroun Al-Raschid; written about the beginning of the 9th century A.D.

A manuscript copy of this work, dated 743 A.H., or 1342 A.D., is preserved in the Bodleian Library, Oxford, and it is surmised to be the earliest copy in existence.

A translation of it was made from the Arabic into English by Fredric Rosen, and published in 1831.

The author, in his preface, states that:—

"Encouraged by the Imam Al Mahmun, Commander of the Faithful, etc., he was induced to compose a short work on calculating, by the rules of completion and reduction, confining it to what is most useful in arithmetic, such as men constantly require in cases of inheritance, legacies, partition law suits, and trade, and in all their dealings with one another, or where the measuring of lands, the digging of canals, geometrical computation, and other objects of various sorts and kinds are concerned."

The design of the work does not extend beyond questions requiring for their solution either simple or quadratic equations, and these are solved by the same rules as those employed in the treatise of Diophantus; but it is not probable that Ben Musa borrowed anything from that work, for it was not till the middle of the fourth century of the Hejeira (about 960 A.D.) that the treatise of Diophantus was translated into Arabic by Abul Wafa Buzani.

Mr. Rosen was of opinion that the Arabs received "their first knowledge of algebra from the Hindus, who furnished them with their decimal notation of numerals, and also with various important points of mathematical and astronomical information ;" but he adds "as to the subject matter of Ben Musa's performance, he seems to have been independent of them in the manner of digesting and treating it ; at least the method he follows in expounding his rules as well as in showing their application, differs considerably from Hindu writers."

It was a matter of much importance to ascertain the degree in which the Arabians were indebted to the Hindus for the improvement made by them in mathematics and astronomy, at the earliest period in which the sciences were cultivated by them. Colebrooke entered upon an investigation of this question,* and gathered to-

* It is stated in the preface to the astronomical tables of Ben-al-Adami, published by his continuator, Al Casem in 920 A.D., that in the reign of the second Abasside Kalif Almansur, in 773 A.D., "An Indian astronomer, well versed in the science he professed, visited the court of the Kalif, bringing with him tables of the planets according to the mean motions, with observations relative to both Solar and Lunar Eclipses, and the ascension of the signs ; taken, as he affirmed, from tables computed by an Indian Prince. The Kalif embracing the opportunity thus presented to him, commanded the book to be published for a guide to the Arabians in matters pertaining to the stars." The task devolved on Muhammed Ben Ibrahim Alfazari, whose version is known to astronomers by the name of the greater Sind-Hind (Arabic of hither and remoter India). It signifies, according to the same author, Ben-al-Adami, the revolving ages. Colebrooke supposes the word may have been Siddhanta or Indu-Siddhanta, and appears to have been that which is contained in the Brahma Siddhanta. It is cited by the astrologer of Balkh Abu Mashar, but he does not specify which of the Indian systems he is citing. But it is distinctly affirmed by later Arabian writers, that only one of the three Indian doctrines of astronomy was understood by the Arabs.—*Colebrooke Essays, Vol. II., p.* 504.

Colebrooke was of opinion that the Sind-Hind was a copy of the revised Brahma Siddhanta of Brahmegupta, and that the fact was deducible from the number of elapsed days between the beginning of the planetary motions and the commencement of the present age of the world according

gether all the information he could find relating to it in the writings of Arabic authors and historians, and the evidence which he brings to bear on the subject appears to prove that during the reigns of the four Abasside Caliphs of Bagdad, Al Mamun, Haroun

---

to the Indian reckoning as it is quoted by Abu Mashar (an astrologer of Balkh), and which agrees precisely with Brahmegupta.—*Colebrooke Essays, Vol. II., p.* 505.

"The work of Alfazari, taken from the Hindu astronomy, continued to be in general use among the Mahomedans until the time of Almamun, for whom it was epitomised by Mohamed Ben Musa Al Khuwarezmi; and his abridgment was thenceforward known by the title of the less Sind-Hind. It appears to have been executed for the satisfaction of Almamun, before this prince's accession to the Caliphate, which took place early in the 3rd century of the Hejira and 9th A.D."—*Colebrooke Essays, Vol. II., p.* 509.

The author of the Tarikhul-Nucama, a writer of the 12th century, 595 A.H., 1198 A.D., quoted by Casiri, observes that "owing to the distance of countries and impediments to intercourse" scarcely any of the writings of the Hindus had reached the Arabians. There are reckoned, he adds, "three celebrated systems (Mazhab) of astronomy; one only of which has been brought to us, namely, the Sind-Hind, which most of the learned Muhamedans have followed." After naming the authors of astronomical tables founded on that basis and assigning the interpretation of the Indian title and quoting the authority of Ben Adami, the compiler of the latest tables mentioned by him, he goes on to say, "that of the Indian sciences no other communications have been received by us but a treatise on music, of which the title in Hindi is Biyaphar, and the signification of that title (fruit of knowledge), the work entitled Calilah and Damanah, upon ethics; and a book of numerical computations which Abu Tafar Muhamed Ben Musa Al Khuwarezmi amplified (basat), and which is a most expeditious and concise method and testifies the ingenuity and acuteness of the Hindus."

The book here noticed as a treatise on ethics is the well-known collection of fables of Pilpai or Bidpai (Sans Vaidyapriya), and was translated from the Pehleir version into Arabic by command of the same Abasside Khalip Almansur, who caused an astronomical treatise to be translated into Arabic.

"The Arabs, however, had other translations from Indian writers, several on Medicine and Materia Medica, another on poisons, and numerous others."—*Colebrooke Essays, Vol. II., pp.* 510–511.

Al Raschid, Al Mamun, and Al Motaded, during a period of about 150 years, from 754 to 904 A.D. the greatest eagerness prevailed to acquire the scientific knowledge of the Hindus and the Greeks, Learned Arabians were employed in translating into Arabic, works that were best known, the Geometry of Euclid, the Brahma Siddhanta of the Hindus, the Almagest of Ptolemy, the Algebra of Diophantus, with various works on music, medicine, etc., from the Sanscrit.

An Indian astronomer was invited to the court of Al Mansur, to give instruction in the Indian astronomy, from which tables were formed. The Indian system was then adopted by the Arabs, and the name Sind-Hind was given to one of the Indian works with which they became best acquainted, and which, according to Colebrooke, appears to have been Brahmegupta's Siddhanta. This book. by command of the Caliph, was used as a guide to the Arabians in matters pertaining to the stars.

Colebrooke concludes his examination as follows:—

"From all these facts, joined with other circumstances to be noticed in progress of this note, it is inferred: First, that the acquaintance of the Arabs with Hindu astronomy, is traced to the middle of the second century of the Hejira in the reign of Al Mansur, upon authority of Arabian historians citing that of the preface of ancient astronomical tables (622 + 150) A.D.; while their knowledge of the Greek astronomy does not appear to have commenced until the subsequent reign of Haroun Alraschid, when a translation of the Almagest is said to have been executed under the auspices of the Barmacide Yahya Ben Khalled, by Abu, Hiau and Salarna employed for the purpose.

"Secondly, that they were become conversant in the Indian method of numerical computation within the second century; that is, before the beginning of Almamun, whose accession to the Caliphate took place in 205 H. (827 A.D.)

"Thirdly, that the first treatise on algebra in Arabic was published in his reign; but their acquaintance with the work of Diophantus is not traced by any historical facts collected from their writings to a period anterior to the middle of the fourth century of the Hejira (972 A.D.), when Abrilwafa Buzjani flourished.

"Fourthly, that Muhamed Ben Muza Khuwarezmi, the same Arabic author, who, in the time of Almamun and before his accession, abridged an earlier work taken from the Hindus, and who published a treatise on the Indian method of numerical computation, is the first, also, who furnished the Arabs with a knowledge of algebra, upon which he expressly wrote.

"A treatise on algebra bearing his name, it may here be remarked, was in the hands of the Italian algebraists, translated into the Italian language not long after the introduction of the science into that country by Leonardo, of Pisa. It appears to have been seen at a later period both by Cardan and by Bombelli. No manuscript of that version is, however, now extant; or, at least, known to be so."

The treatise on arithmetic and Algebra entitled "Liber Abbaci," by Leonardo, the son of Bonacci, of Pisa, was published in 1202 A.D. In the account which he gives of himself in the preface of his work, he says that he travelled into Egypt, Barbary, Syria, Greece, and Sicily; that being in his youth at Bugia in Barbary, where his father Bonacci, held an employment of scribe at the custom house, by appointment from Pisa (for Pisan merchants resorting thither), he was there grounded in the Indian method of accounting by nine numerals. Further, that finding it more commodious, and far preferable to that used in other countries visited by him, he prosecuted the study, and, with some additions of his own, and taking some things from Euclid's Geometry, he undertook the composition of the treatise in question, that "the Latin race might no longer be found deficient in the complete knowledge of that method of computation," and he professes to have taught the complete doctrine of numbers according to the Indian method.

The treatise on algebra by Diophantus, before referred to as having been translated from the Greek into Arabic, in the reign of the Caliph Al Motaded, about A.D. 900, although now well known, was apparently unknown to European mathematicians before the time of Regiomentanus. He (in the preface to the "Elements of Astronomy," of the Arabian astronomer, Alfragan, whose name is derived from the place of his birth, Fergan in Sogdiana or Samarcand, and who flourished about 800 A.D.), informs us that Diophantus wrote 13 books on arithmetic and algebra, which are still preserved in the Vatican Library. Bombelli, in the preface to his "Algebra" (1572 A.D.), says that there were only six of these books then in the library, and that he and another were engaged in a translation of them. These six books have been published in Greek and Latin at different later times. Those particularly mentioned are two editions: one by Bachet, Paris, 1621; the other with notes by Fermat, Toulouse, 1670.

As a science little known to the Greeks of later times, the Diophantine Analysis was a subject on which, according to Suidas, the celebrated Hypatia, in her capacity of President of the Alexandrian School of Philosophy, lectured before that society, as her father, the mathematician Theon, did also in the same office on the Syntaxis of Ptolemy.

Now, in the year 415 A.D., Hypatia was brutally murdered by a mob of monks in an outbreak against the Governor of Alexandria, Orestes. For 1,000 years afterwards the work of Diophantus does not appear to have been known, except in name, to either Greeks or Romans, although it was known to the Arabs, and appreciated by them in the reigns of the Abasside Caliphs of Bagdad, soon after the Arabs were in possession of Alexandria.

The age when it was supposed to have been written is variously stated by different writers, and it was supposed to have had its origin in Alexandria.

Abulfaraj considers Diophantus to have been a contemporary of the Emperor Julian about 365 A.D.

Other writers suppose the date to have been 150 A.D. Bachet conjectures that the age in which he flourished was about the time of Nero, 54 A.D. Cossali, in his "Origine dell Algebra," was of opinion that he lived about 200 B.C.

Amid so much conjecture and uncertainty with reference to the origin of this work, a suggestion may be permitted by way of explanation regarding it, namely, that the book was a translation from some ancient original manuscript, one out of the numerous rolls then in the library, which had been brought from the East, the spoils of war in the Asiatic Campaign of Alexander. This supposition would appear to receive support from the meaning of the Greek word Diophantus, as a title to the book, it would signify "Explained by the Gods." Now, most Indian works on science are supposed to have a divine origin, but this work differs in some respects from known Indian works on algebra, as will be explained hereafter. It may possibly have had a Persian or a Chaldaic origin.

That an Asiatic origin is most probable, derives evidence from the manner in which the Alexandrian Library was formed and received its increase.

We are told that Ptolemy Soter, the favourite General of Alexander, was a great lover of literature and science. He had a passion for the collection of manuscripts, and had ample opportunities for the indulgence of this favourite pursuit, in the campaign in Asia, the literary wealth of which he acquired, and the manuscripts of which he collected as the spoil of war, and carried away from its palaces and temples.

When he became fully settled in his sovereignty as King of the Egyptian province, to which he had succeeded after the death of Alexander, he devoted much time to the formation of a library. This was undertaken at the suggestion of Demetrius Phalierus, who had taken refuge in Alexandria on his flight from Athens, where he had been Governor, being received with great hospitality by King Ptolemy. This library was that which, under his successors, Philadelphus and Uergetes, who inherited his father's love of the science, was increased to about 400,000 volumes, among which were valuable and curious manuscripts from most countries then known.

It is said of Uergetes that he adopted a most unscrupulous method of adding to the library, as, for instance, that he seized books imported into Egypt from neighbouring countries, and, having caused them to be copied, returned the copies to the owners, keeping the originals for the library.

There must have been many ancient manuscripts in this vast collection not written in the Greek language, so Ptolemy adopted a course which was best calculated to make him acquainted with their contents.

He made his court an asylum for learned and talented men, who, from war or persecution, having been driven from their homes, and being received and established under his own protection, were treated with munificence and liberality, and the doctrines they professed listened to with toleration.

The members forming this great society lived together and partook of the common bounty of the Sovereign. They formed four principal schools, of which the first consisted of critics and commentators, the second of mathematicians, the third of practical astronomers, and the fourth was a school of medicine and anatomy, of which last, we are informed, one professor, named Herophilus, dissected 600 men!

By his example in sharing their labours and taking part in their discussions on philosophical subjects, he excited emulation and aroused a spirit of enquiry, which raised the Alexandrian School to the highest distinction in literature and science.

For about 300 years before the conquest of Egypt by the Romans this School flourished and became famous by reason of the distinguished philosophers who were members of it.

Among the mathematicians and astronomers of this period, whose names have descended to us, and which are deservedly honoured at the present day, are those of Euclid (280 B.C.), Aristarchus, Eratosthenes (240 B.C.), Conon, and, according to Troclus, Archimedes (250 B.C.), who is said to have studied under Conon, in the reigns of Philadelphus and Uergetes. About the same time there flourished the geometricians Apollonius and Nicomedes, and a little later the eminent astronomer Hipparchus (135 B.C.), to whom Ptolemy, the astronomer (70 A.D.), was so greatly indebted in the compilation of his great work called the "Syntaxis"—the Almagest of the Arabians.

The principles of mathematics embodied in his various works by Euclid were, before his time, taught by Plato (390 B.C.) and by Pythagoras (550 B.C.), and, doubtless, to some extent by other more ancient writers.

The great merit of Euclid was, that he reduced to order the fundamental principles delivered by the earlier writers, and by his admirable arrangement of them formed that system of logic, which, step by step, carried conviction of their truths to the mind, by irrefragible demonstration—a system of reasoning which has never been surpassed, and which, in his "Elements of Geometry," still holds its own in the schools of the present day.

The aid which must have been afforded by the library to the philosophers of the Alexandrian School is incalculable. To state the degree in which the more ancient sciences were embodied in the writings of this period is impossible. The subject has, in a great measure, been avoided, and modern writers have been content to ascribe to eminent men of the Alexandrian School discoveries which were, in fact, made long before that School was established.

It cannot be doubted that, stored in the library, were many manuscripts, containing the wisdom of ancient Eastern nations, of which there is now no record, but which must have been translated and embodied in the works of Greek authors at about this period.

In the words of Laplace:—

"Such have been the vicissitudes of human affairs, that great nations, whose names are hardly known in history, have disappeared from the soil which they inhabited; their annals, their language, and even their cities have been obliterated, and nothing is left of their science and industry but a confused tradition and some scattered ruins of doubtful and uncertain origin."

About the years 1587 and 1634 A.D., Akber, the Emperor of India, caused translations to be made from the Sanscrit into the Persian language, of the Lilawatee and the Vija Ganita, treatises on arithmetic and algebra of the Hindu mathematician, Bhascara Acharya, which have already been referred to in a previous section of this work. The first of these was translated by Abul Fazel, the confidential minister of Akber, and the second by Utta Ulla Rushudee. They are both compilations from ancient Hindu works, connected with numbers, geometry, and mensuration.

At the end of the last century partial translations, of these works of Bhascara, from the Persian into English, were made by Mr. Davis and Mr. Reuben Burrows, and a complete translation was made from the Persian by Mr. Edward Strachey, of the Indian Civil Service. In 1817, Mr. Colebrooke published his translation of these treatises directly from Sanscrit versions, and he added various notes and the commentaries of other Hindu writers.

In 1796, Dr. Hutton, in his "History of Algebra," gave a short description of these two works of Bhascara, and, as a result of the investigation which he made with reference to the origin of algebra, he expresses his opinion as follows:—

"From a comparison of the algebra of the Arabians and the Greeks and that of the modern Europeans, with the Persian translation of the Vija Ganita and the Lilawatee, it would appear that the algebra of the Arabs is quite different from that of Diophantus, and not taken the one from the other; that if the Arabs did learn from the Indians, they did not borrow largely from them.

"That the Persian translations of the Vija Ganita and Lilawatee contain principles which are sufficient for the solution of any proposition in the Arabian or in the Diophantine algebra; that these translations contain propositions which are not to be solved on any principles that could be supplied by the Arabian or the Diophantine algebras; and that the Hindus were further advanced in some branches of this science than the modern Europeans, with all their improvements, till the middle of the 18th century."

He further remarks:—

"General methods for the solution of indeterminate problems are found in the fourth and fifth chapters, which differ much from Diophantus' work. Hindu Algebra contains, in Mr. Strachey's opinion, which is highly probable, a great deal of knowledge and skill, which the Greeks had not, such as the use of an indefinite number of unknown quantities and the use of arbitrary marks to express them; a good arithmetic of Surds; a perfect theory of indeterminate problems of the first degree; a very extensive and general knowledge of those of the second degree; a perfect acquaintance with quadratic equations, etc.

Hutton continues: "the arrangement and manner of the two works are as different as their substance; the one constitutes a regular body of science, the other does not; the Vija Ganita is quite connected and well digested, and abounds in general rules, which suppose great learning; the rules are illustrated by examples, and the solutions are performed with skill.

"Diophantus, though not entirely without method, gives very few general propositions, being chiefly remarkable for the dexterity and ingenuity with which he makes assumptions for the simple solution of his questions. The former teaches algebra as a science, by treating it systematically; the latter sharpens the wit by solving a variety of abstruse and complicated problems."

The solution of certain problems, by the application of Algebra to Geometry, is remarked upon by Mr. Strachey.

Some of these have names peculiar to themselves. Thus, the figure designated as the "bride's chair," "the wedding chair," is a square made up of the four right-angled triangles, which are equal to twice the rectangle of their sides, together with the small square which is the square of the difference of the sides. Mr. Strachey was of opinion that the Hindus were well acquainted with most of the propositions in Euclid's elements.

It is easy to see from the following original proposition in the Hindu work, that the Pythagorean theorem is intended to be understood.

"The square of the hypotenuse of every right-angled triangle is equal to twice the rectangle of the two sides containing the right angle with the square of the difference of those sides."

It is evident that if x and y be the sides and z the hypotenuse, the proposition makes—

$$z^2 = 2xy + (x - y)^2 \text{ and obviously}$$
$$= x^2 + y^2.$$

The geometrical proof being that the two rectangles are equal to the four right-angled triangles containing the sides; and that, with the square of the difference of the sides, they together make up the figure of the bride's chair.

The name of the figure, it is conjectured, has been suggested by its resemblance to the tonjon or palanquin, in which it was usual for the bride to be carried to her husband's house.

Dr. Hutton was of opinion, from the many questions about right-angled triangles worked algebraically, that it was probably in India where Pythagoras acquired his mathematical knowledge which he carried back with him and taught to his countrymen. With reference to the algebra of the Greeks, he says:—

"It is doubtful if the Greeks had any other algebra than that of Diophantus, and it may be worthy of remark that at Alexandria he may have had access to Indian literature."

Or, what is more likely, that, as before suggested, the treatise entitled Diophantus was of a more Eastern origin, supposed to be divine, as the name implies. Since the Greek translation was found in the library in later times, with no explanation of its true origin, the author has been supposed to be some Greek named Diophantus!

It has been before suggested that the earliest form of the sun dial was the shadow cast on the horizontal plane by the tent-pole of the Nomadic tender of cattle, wandering over the great level plains of Asia. The changes which the shadow underwent during the day and throughout the year must have been always noticed by the inhabitants of the high lands or steppes of Asia long before the right-angled triangle formed by the upright pole, the shadow and the solar ray joining their extremities, became a subject of much attention. By all, it would have been seen that the shadow, growing less from sunrise to midday, increased again from noon to sunset. They noticed during the year that the midday shadow varied in length, being shortest when the sun was in the Summer Solstice, and longest at the Winter Solstice.

Their occupation of tending flocks made these phenomena familiar to them, and the points of the horizon at which the sun would rise would be noticed by trees or other marks, or by the direction of the shadow produced backwards, and the points of sunrise would be seen to change throughout the year, oscillating, as it were, to extreme points North and South of the East points. Thus, they may have been long familiar with these appearances before measurements of the upright, the shadow, and the line between their extremities were made.

The innumerable changes which the form of the triangle and the length of the shadow underwent at first could not fail to be a subject of great perplexity; but it is most probable that these measurements led first to the numerical discovery that the squares on the sides of the right-angled triangle added together are equal to the square described on the hypotenuse.

Now, it is a prominent feature of Hindu mathematics in their algebra, geometry, and astronomy, that the numerical properties of the right-angled triangle are principally employed in the solution of problems. General theorems were framed for special purposes. Theorems, relating both to abstract and concrete numbers, were invented, and applied to problems, the solutions required being sometimes in integers, and at other times, in a more general rational form, integral or fractional.

The solution of problems in concrete numbers was such as had relation to the requirements of the age in which the rules were formed; measurements of land, contents in the excavation of canals, tanks, etc.; the number of bricks in piles of different forms; contents of conical mounds of grain and stacks; sawing of timber; interest of money; purchase and sale, etc.

A few examples are here inserted, extracted from the Lilavati and the Vija Ganita, and expressed in the modern form, in order to illustrate some of the methods employed in those treatises:—

Ex. If x and y be sides and z the hypotenuse, then for the solution of the indeterminate equation—

$$x^2 + y^2 = z^2, \text{ all in integers.}$$

Let m and n be any two arbitrary integers, m greater than n, we have for solutions—

$$x = 2\,m\,n.$$
$$y = m^2 - n^2.$$
$$z = m^2 + n^2.$$

For solutions generally in rational numbers, integral or fractional, of the equation, $x^2 + y^2 = z^2$.

When one of the sides x is given—

Assume $y = \frac{2\,m}{m^2-1}\;.x$ (m being any arbitrary integer).

And $z = \frac{m^2+1}{m^2-1}\;.x.$

Otherwise—

$$\text{Assume } y = \tfrac{1}{2}\left(\frac{x^2}{m} - m\right)$$

$$z = \tfrac{1}{2}\left(\frac{x^2}{m} + m\right).$$

When the hypotenuse z is given—

$$\text{Assume } x = \frac{2\,m\,z}{m^2+1}$$

$$y = \frac{m^2-1}{m^2+1} \cdot z$$

Again, Ex. 60, for the solution of the equation, $x^2\,y^2 - 1 = z^2$.

$$\text{let } x = \frac{1}{2\,m} + m + 1, \text{ where } m \text{ is arbitrary.}$$

$$y = \frac{1}{2\,m} + m - 1, \text{ and } z = \frac{1}{4\,m^2} - m^2.$$

$$\therefore\ x^2\,y^2 = \left\{\left(\frac{1}{2\,m} + m\right)^2 - 1\right\}^2$$

$$\therefore\ x^2\,y^2 - 1 = \left(\frac{1}{4\,m^2} - m^2\right)^2 = z^2.$$

Ex. (61.) For the solution of the equation $x^2 + y^2 - 1 = z^2$

Assume $x = 8\,m^4 + 1$, where m is arbitrary

And $y = 8\,m^3$

$z = 4\,m^2\,(2\,m^2 + 1)$

consequently $x^2 + y^2 - 1 = 16\,m^4\,(2\,m^2 + 1)^2 = z^2$.

(201) Rule.—When the diameter of a circle is multiplied by three thousand nine hundred and twenty-seven, and divided by twelve hundred and fifty, the quotient is the near circumference; or, multiplied by twenty-two, and divided by seven, it is the gross circumference. Thus:—

$$\text{Near circumference} = \frac{3927}{1250} \times d = 3{\cdot}1416 \times d.$$

$$\text{Gross circumference} = \frac{22}{7} \times d.$$

(203) Rule.—In a circle, a quarter of the diameter multiplied by the circumference, is the area.

That, multiplied by four, is the net all round the ball. This content of the surface of the sphere, multiplied by the diameter and divided by six, is the precise solid, termed cubic, content within the sphere. Thus:—

Area of a circle $= 3{\cdot}1416 \times \frac{d^2}{4}$.

Surface of a sphere = four times area of a great circle.

Contents of a sphere $= 3{\cdot}1416 \times \frac{d^3}{6}$.

Ganesa shows how the area of a circle is the product of the semi-circumference and the semi-diameter, thus:—Dividing the circle into two equal parts, cut the content of each into any number of angular spaces, and expand it so that the circumference becomes a straight line. Thus:—

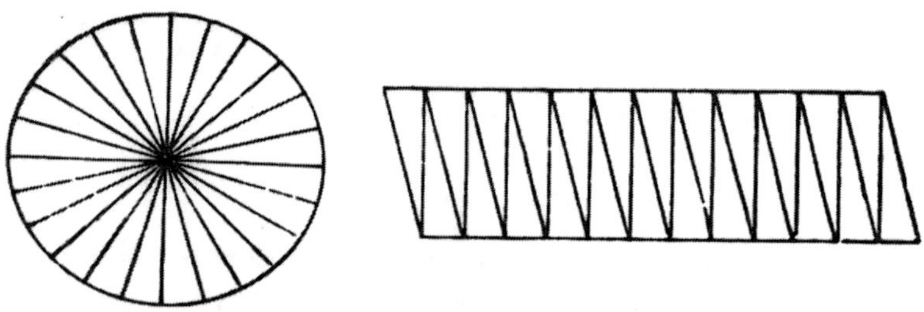

Then let the two portions approach so that the sharp angular spaces of the one may enter into the similar intermediate vacant spaces of the other, thus constituting an oblong, of which the semi-diameter is one side and half the circumference the other. The product of their multiplication is the area.

Ganesa then proceeds to demonstrate the rule for the solid content of the sphere, thus:—

Suppose the sphere divided into as many little pyramids or long needles, with an acute tip and square base, as is the number by which the surface is measured, and in length (height) equal to half the diameter of the sphere; the base of each pyramid is an unit of the scale by which the dimensions of the surface are reckoned; and the altitude being a semi-diameter, one-third of the product of their multiplication is the content; for the needle shaped excavation is one-third of a regular equilateral excavation, as shown in § 221. Therefore (unit taken into) a sixth part of the diameter is the content of one such pyramidical portion; and that, multiplied by the surface, gives the solid content of the sphere.

If there be one thing which distinguishes Hindu astronomy from the modern more than another, it is in the assumed radius of a circle.

The circumference of any circle being divided into 360 degrees or 21,600 minutes of arc, then, if the radius be supposed to be a flexible line wrapped along the circumference, it will cover of these divisions, 57° 17′ 44″ 48‴, &c.

This, in every circle, is an arc equal in length to its radius. If it be reduced to minutes it becomes very nearly 3437¾ minutes.

The nearest integer to this mixed number is 3438. It differs by little more than 15 seconds from the actual length, and this number is assumed to be the radius of any circle in its own minutes of arc; in ordinary cases the difference would be scarcely appreciable.

Thus, every circle furnishes its own scale for reckoning straight lines, in minutes of arc. It is in this scale that the Hindu table of sines and versed sines is formed, in nearest integers of minutes, for 24 arcs of a quadrant, the arcs differing from each other by 225′, or 3° 45′.

The Hindus divide the number of minutes in a semi-circle, 10,800, by the minutes in the radius, 3438, and obtain 3.14136, as the ratio of the circumference to the diameter of a circle, differing from 3.14159 (commonly assumed by us) by .00013 only! This (the Hindu) may have been one of the most ancient methods of calculating the circumference of a circle from its radius.

CHAPTER VIII.

ASTRONOMICAL INSTRUMENTS.

The principal instrument (if it may be so called) used by the Hindu astronomers was the Moon, which, from the rapidity of her motion, and the known places of the fixed stars on each side of her path, was an efficient means of determining the positions and motions of the planets by referring them directly, as she passed, to the nearest Yoga-taras of her course.

During the day observations were made, of the sun's altitude, and amplitude, and for the times of observation, by means of the shadow cast by the vertical gnomon of a dial, and differences of time from sunrise, estimated for astrological purposes.

Bhaskara gives a brief detail of a few astronomical instruments which were in use in his time, but he says—" of all instruments, it is Ingenuity which is the best."

Colebrooke, in his essay on the Indian and Arabian divisions of the Zodiac (Vol. 9 of the "Asiatic Researches"), says:—

" The manner of observing the places of the stars is not explained in the original works cited. The Surya Siddhanta only hints briefly that the astronomer should frame a sphere and examine the apparent longitude and latitude. The commentators, Ranganatha and Bhudhara, remarking on the passage, describe the manner of the observation, and the same description occurs, with little variation, in commentaries on the Siddhanta Siromani. They direct the spherical instrument Golayantra to be constructed according to the instructions contained in a subsequent part of the text. This is precisely an armillary sphere. An additional circle, graduated for degrees and minutes, is directed to be suspended on the pins of the axis as pivots. It is named Vedhavalaya, or intersecting circle, and appears to be a circle of declination. After noticing this addition to the instrument, the instructions proceed to the rectifying of the Golayantra, or armillary sphere, which is to be placed so that the axis shall point to the pole, and the horizon be true by a water level.

"The instrument being thus placed, the observer is instructed to look at the star Revati, through a sight fitted to an orifice, at the centre of the sphere; and, having found the star, to adjust, by it, the end of the sign Pisces on the Ecliptic. The observer is then to look through the sight, at the Yoga star of Aswini, or of some other proposed object, and to bring the moveable circle of declination over it. The distance in degrees from the intersection of this circle and Ecliptic to the end of Mina or Pisces, is its longitude, dhruvaca, in degrees; and the number of degrees on the moveable circle of declination, from the same intersection to the place of the star, is its latitude, Vickshepa, North or South."

From this description it will be seen that when the Yoga-tara, or principal star of an Asterism, is in the Meridian of any place, then, at the same moment, a point of the Ecliptic is determined, namely, the point of intersection of the Ecliptic with the Meridian; this point is called by the Hindus the Kranti Pata, the intersecting point, of the Ecliptic, with a circle of declination. The arc of the circle of declination from this point to the star is called the apparent latitude, and the arc of the Ecliptic, from the same point to the beginning of the Asterism Aswini, or to the first point of Mesha, is the star's apparent longitude. The observations necessary to find these co-ordinates, for the place of a star, were not difficult to accomplish, especially when Hindu astronomers had different means for effecting the same object.

The Armillary Sphere, however, was of a nature too complicated to be used as an instrument for making accurate observations, and was rather for the purposes of explanation, and of giving instruction on the numerous circles and motions of the several spheres of which it was composed.

It consisted of at least three separate spheres, on the same polar axis, or Dhruva-Yasti. First: A fixed celestial sphere named the Khagola, composed of circles for a given latitude, such as the horizon, the equinoctial, the meridian the prime vertical, the six o'clock hour circle, the vertical circles through the N.E. and N.W. points of the horizon; the names of these circles respectively in this order are the Kshitija, the Nadi-Valaya (marked with 60 Ghatikas), the Yamy-ottara-Yritta, the Samamandala, the Unmandala, the two Kona-Vrittas, with other circles, which remain always the same for the same place. Besides these fixed circles, a moveable altitude and azimuth circle is attached, by a pair of pins, to the zenith and nadir points of the Khagola, for showing the altitude or azimuth of any star. The horizon being divided in degrees, either from the Meridian line or from the East and West points. Secondly: Moveable within and round the axis of the Khagola was the starry sphere named the Bhagola, which comprised the Ecliptic, with the paths also of the moon and planets, named Kshepa-Vritta, the circles of declination, or Kranti, the diurnal circles called Ahoratra-Vrittas, the azimuth circle through the Nouagesimal point, is called the Drikshepa-Vritta. The Bhagola is supported within the Khagola by means of two supporting circles called the Adhara-Vrittas, corresponding with the Meridian and horizon of the Khagola. Thirdly: On the axis of the Khagola produced, a third sphere is supported. It is called the Driggola, or double sphere, which is a system in which the circles of the Khagola are mixed with those of the Bhagola. The Khagola and Driggola remain fixed, while the Bhagola alone revolves.

Bhaskara also gives a brief description of several other instruments, among which are the Nadi-Valaya, a circle representing the Equinoctial divided into Ghatikas, and on it are the positions of the 12 signs, calculated to correspond with their oblique ascensions or risings at the place of observation. It is used in connection with the Khagola, whose axis casts a shadow on the circle, and is, in fact, an equatorial dial, the Ghatika being $\frac{2}{5}$ of an hour.

An instrument for time, the Ghati or Clepshydra, made of copper, is like the lower half of a water pot. A hole is made in its bottom, and when placed on a vessel of water, the size of the hole is adjusted so that it will sink to the bottom 24 or 60 times in a day, hence the name Ghati.

The Chakra, or circle, marked on its circumference with 360°, is suspended by a string, the beginning of the divisions being at the lowest point. At the centre is a thin axis perpendicular to its plane. When the instrument is turned so that its plane is coincident with a vertical circle passing through the sun, the shadow of the axis is thrown on some division of the circumference and the arc between this point and lowest point, the zero of the divisions, measures the *zenith distance* or co-altitude of the sun. It is also used for finding the longitude of a planet; for if the instrument be inclined, and held or fixed so that any two of the stars Regulus, δ Cancri, ζ Piscium, or λ Aquarii, appear to touch the circumference, the plane of the circle will coincide with the plane of the Ecliptic, since these stars have no latitude. (Spica, whose latitude is inconsiderable, 2° S., and other stars near the Ecliptic, would appear also to touch the circumference.) The latitude of a planet, also, which is in general very small, has its orbit nearly in the same plane with that of the Ecliptic. Looking, then, through a sight at the zero point of the circle, so that the planet appears opposite the axis, the position of the circle then remaining fixed, the eye is moved along the lower part of the circumference, so that any one of the above stars is seen opposite the axis, the arc between the two positions of the eye is the difference of longitude between the planet and the star; but the longitude of the star being known, that of the planet will also be known.*

---

* In the "Philosophical Transactions," Vol. LXVII., p. 598, are drawings of astronomical instruments found in an observatory at Benares by Sir Robert Barker, who visited it in 1772 A.D., these were of large dimensions and constructed with great skill and ingenuity.

The traditionary account is that the observatory was erected by the Emperor Akber.

The modern Armillary Sphere was of a less complicated nature, as will be understood from the accompanying diagram, which has been copied from a plate dated A.D. 1720.

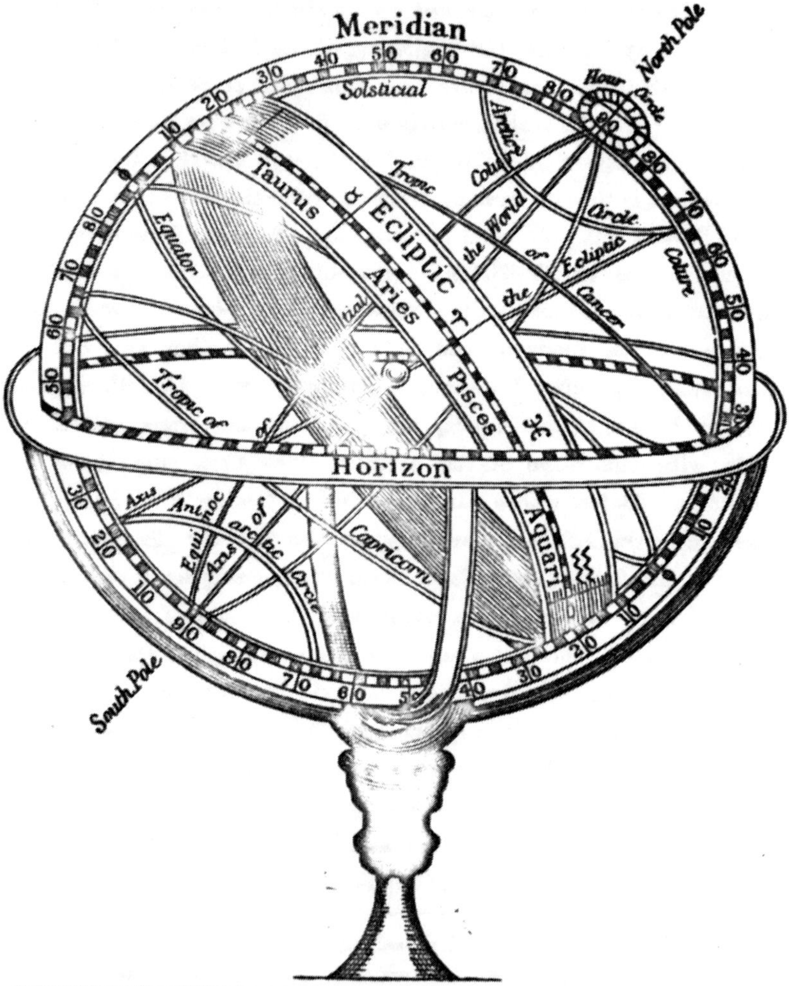

"P. Tiessenthaler describes in a cursory manner two observatories furnished with instruments of extraordinary magnitude at Jeypoor and Oujein, in the country of Malwa, but these are said to be modern structures."—*Robertson, p.* 438.

The ball at the centre represents the earth, with lines on its surface corresponding with the circles of the Celestial Sphere, Meridians, Parallels, &c., as also the configurations of seas, countries, &c.

The Zodiac was a band extending to about six degrees on each side of the Ecliptic, within which are confined the traces of the Moon's path, in spiral convolutions and the orbits of the five planets, the field of all their encounters with each other, their conjunctions, occultations and eclipses—natural occurrences which gave rise to the wild fancies of the astrologer, and invested him with such terrible powers over the fears and fanaticism of the superstitious, in which he still holds sway over many millions of Asiatic nations.

The Armillary Sphere held its place as an astronomical instrument till near the beginning of the present century, and was used for the solution of astronomical problems, until the more accurate instruments were introduced for observing the passages of celestial bodies across the meridian.

Near the middle of the last century an Armillary Sphere was constructed by Dr. Long, Master of Pembroke College, Cambridge, and Lowndes Professor in that University. This instrument was "18 feet in diameter, and would contain more than 30 persons within it, to view as from a centre the representation of the Celestial Spheres. The whole apparatus was so contrived that it could be turned round with as little labour as is employed to wind up a common jack."

CHAPTER IX.

SOME EARLY HINDU ASTRONOMERS AND OBSERVATIONS.

[*Circ. B.C.* 1590—945.]

In a work dealing with any system of science such as astronomy, as studied and practised by a nation, it is, of course, expected that some account should be given of the men who founded it, or, at any rate, devoted their attention to the subject during its infancy.

The history of the age in which they lived would, in such account, be necessarily referred to as a component part of it, and as establishing their place in that history. Hitherto the writer has been obliged to refer to the astronomers of the early periods of the Aryan immigration to India in purely generic terms. Although, from a contemplation of the Hindu system of astronomy, and from an examination of the Hindu works on that science (some of which are intended to be hereafter considered), it is absolutely certain that there were men of great genius, living in those distant ages, having attainments and abilities far beyond those of astronomers of a later date, yet their names are lost, except so far as we can vaguely connect them with the "Rishis" or "Munis," the sages to whom reference has been casually made.

In later writings of Hindu astronomers, however, there are several names specifically mentioned connected with ancient astronomical observations said to have been made by them, which observations assist the investigator of facts in finding a place in history for the bearers of those names. It is to be regretted that the history of the Hindu race, relating to the periods in which the ancient astronomers lived, is so wrapped up in their mythological cosmogony and fabled legends, and so connected with their doubtful chronology, that no great reliance can be placed upon the conjectures and conclusions made and arrived at, from Hindu writings and traditions, by various investigators. It is, nevertheless, interesting to trace the circumstances leading to such conjectures, and, at any rate, to point to the men who undoubtedly existed, and as undoubtedly assisted in establishing, if they did not themselves originate, the system of Hindu astronomy which it is the object of this work to discuss.

In the first place, it is necessary to make a marked distinction between those circumstances which are derived from a consideration of the great epic poems of the Hindus, the Mahabharata and the Ramayana, and even of their religious scriptures, such as the Vedas and the Institutes of Menu, and the circumstances connected with purely astronomical deductions. Still more is it necessary to bear in mind, as a separate set of circumstances, those derived from the admitted fables and mythology of the Hindus.

There are, nevertheless, to be gathered from all these data the certain facts that (at a period to be located in history) there lived and flourished in India two royal dynasties, the one styled "The Children of the Sun," the descendants of whose family are supposed to have reigned in the city of Ayodhya, and "The Children of the Moon," who reigned in Pratishthana, or Vitora. It is here supposed that they were so styled according to the manner in which they reckoned their astronomical time, whether from the use of a Solar or a Lunar Zodiac.

Of the former princes, the name standing out most prominently in Hindu history is that of Rama, the son of Dasaratha, of the Solar race, the hero of the great epic poem of the Hindus called the Ramayana.

The princes of the latter dynasty find a place in the other poem—the Mahabharata—which describes the events of the great war between the Pandus and the Kurus, the successful issue of which was in favour of the Pandus, five brothers of the Lunar race, the chief of whom was Yudhisthira.

I

Contemporary with this prince were the two Indian astronomers, Parasara and Garga. The precise period in which they, and also Yudhisthira, existed, and, therefore, the period of the Mahabharata, is a vexed question, which it is needless in this work to enter into, except in a cursory manner.

This much, however, is most probable, if not certain—that Parasara and Garga were men of astronomical genius, and that Yudhisthira lent his powerful aid in the development of their researches, as evidenced by the activity apparent in the study of the heavens during that remote period.

Captain Wilford, in his "Chronology of the Hindus," says:— "It has been asserted that Parasara (who was contemporary with Yudhisthira) lived about 1180 B.C., in consequence of an observation of the places of the colures. But Mr. Davis, having considered the subject with the minutest attention, authorises me to say that this observation must have been made 1,391 years before the Christian Era. This is also confirmed by a passage from the Parasara Sanhita, in which it is declared that the Udaya, or heliacal rising of Canopus (when at the distance of 13° from the sun, according to the Hindu astronomers), happened in the time of Parasara on the 10th of Cartica; the difference now amounts to 23 days. Having communicated this passage to Mr. Davis, he informed me that it coincided with the observation of the places of the colures in the time of Parasara."

Sir W. Jones found great difficulty in reconciling the fables of the Hindus, so as to obtain probable dates for the times of Yudhisthira and Rama. He says:—"We find Yudhisthira, who reigned confessedly at the close of the brazen age, nine generations older than Rama, before whose birth the silver age is allowed to have ended.

"Paricshit, the great nephew of Yudhisthira, whom he succeeded and who was the grandson of Arjun, is allowed, without controversy, to have reigned in the interval between the brazen and earthen ages, and to have died at the setting in of the Kali Yuga (3102 B.C.)"

According to one hypothesis, Paricshit is placed at 1029 B.C., and by another he would have a probable date of 1717 B.C. On the other hand, a hypothetical date of Rama, according to Hindu tradition, was even so early as 2029 B.C.

From various passages in the Varhi Sanhita, it has been inferred that Varaha Mihira (its author) had made observations on the position of the Solstitial Colure in his time,* and that he compared it with the position it occupied in the time of Parasara—a period when, as stated by that author, the Solstitial points were, the one in the middle of Aslesha, and the other in the beginning of Dhanishtha. Thus, "a passage cited by Bhattotpala, the commentator of Varaha Mihira, corresponds in import to a pasage quoted by Mr. Davis and Sir W. Jones from the third chapter of the Varahi Sanhita." The passage referred to, and translated by Colebrooke, is:—

"When the return of the sun took place from the middle of Aslesha, the tropic was then right. It now takes place from Punarvasu."

From this and other similar passages, it was reckoned by Colebrooke, Sir W. Jones, and Davis, that the time when the Solstitial Colure occupied such a position corresponded to the year 1181 B.C., and that this was the time when Parasara was living.

Reckoning, however, the precession or regression of the Solstice at a mean annual rate of 50", the period at which the Solstice was in the middle of Aslesha would be 1110 B.C., the difference between

---

* From a statement of Varaha, that the solstices in his time, as referred to by him, was one in the first degree of Carcata, and the other in the first of Marcara, the period when he (Varaha) lived was deduced by Sir W. Jones and Bailly to have been 499 of our era. The astronomers of Ujain place the date of Varaha at 505 A.D.; and Colebrooke from the position of the Colures, with respect to Spica Virginis, computed the date to be 472 A.D. The greatest difference between these dates being 33 years, was within the duration of a man's life. Any of these dates might therefore represent the time when Varaha lived.

I 2

the dates being accounted for by the lower rate of precession assumed by Colebrooke and others in computing the time.

The same date (1181 B.C.) would appear to have been adopted by Bentley for the place of the Solstice in the middle of Aslesha, which agrees with certain calculations of his own from other sources.

But he does not accept the opinion that Parasara was then living, giving reasons for supposing that the date when this astronomer, with Yudhisthira and Garga, flourished, and when the war of Bharata took place, was some 600 years later.

A somewhat inferior method of determining astronomical dates than by means of the Colures has sometimes been employed, namely, by computing from rules given by Hindu astronomers regarding the heliacal rising and setting of a particular star. Thus, the rising and setting of Agastya or Canopus appear in India to have been important on account of certain ceremonies to be performed when that star appears, rising with the sun. The rising and setting of this star is referred to by Varaha Mihira and others. Varaha-Mihira says :—"Agastya is visible at Ujjayni when the sun is 7° short of the sign Virgo"; but he afterwards adds, "the star becomes visible when the sun reaches Hasta, and disappears when the sun arrives at Rohini." His commentator remarks that the author has here followed earlier writers, and has quoted Parasara as saying: "When the sun is in Hasta the star rises, and it sets when the sun is in Rohini." Upon this, Colebrooke remarks that it is probable Parasara's rule was framed for the North of India. It will be seen, however, that if the date when the star rose as indicated be established, the date of Parasara's assertion is also established.

Bentley (as before stated) contended that both Colebrooke and Davis were wrong in placing the date of Parasara, and, consequently, of Yudhisthira, at 1181 B.C.; and, from a theory of his own, he calculates the date to have been 575 B.C.

The passage referred to regarding the heliacal rising of Canopus states that "the star Agastya (or Canopus) rises heliacally when the sun enters the Lunar Asterism Hasta, and disappears, or sets, heliacally when the star is in Rohini."

From the times of rising and setting of Canopus thus given, Bentley calculates the latitude of the place of observation, which he finds to be nearly that of Delhi, 28° 38′ N., and for his supposed date, 575 B.C., he finds the longitude of Canopus 68° 47′ 10″, the latitude 76° 8′ 32″ S., the right ascension 81° 43′ 25″, and the declination 52° 58′ 53″ S.

From these data he seeks the longitude of the sun from the Vernal Equinoctial Point at the time, when the star Canopus rose heliacally at Delhi, in the year 575 B.C., which he finds to be 145° 10′ 5″.

To compare this with the observation of Parasara, he ascertains, by reference to Cor Leonis, in the British Catalogue of 1750, the longitude of the beginning of Hasta from the Vernal Equinoctial Point in the year 575 B.C. to be 145° 4′ 12″.

The difference being only 5′ 53″, he concludes to be sufficient proof of the accuracy of the observation of Parasara.

He further remarks that the place of observation was a few miles to the South of Delhi, called Hastina-pura, the seat of government in the time of Yudhisthira, which would make the agreement between the observation and the calculation still more correct.

Bentley's contention that the epoch of Yudhisthira and Parasara was 575 B.C. (and not 1181 B.C., as stated by others) would appear to receive some confirmation from certain further statements of Varaha Mihira, who, in the Varahi Sanhita (472 A.D., see note p. 115), has a chapter expressly on the subject of the supposed motion of the Rishis in Magha.

Colebrooke says he (Varaha) begins by announcing his intention of stating their revolutions conformably with the doctrine of Garga, and proceeds as follows:—"When King Yudhisthira ruled the earth, the Munis were in Magha, and the period of the era of that king is 2526. They remain for a hundred years in each Asterism, being connected with that particular Nacshatra to which, when it rises in the East, the line of their rising is directed."

In this statement of Varaha, if the date could be relied upon as authentic, there is one item which would settle a much-disputed point, namely, the date of the great war between the Solar and Lunar races of the Aryans, or the war between the Pandus and the Kurus, for Yudhisthira was the brother of the four Pandus who were the victors in that war, in the battle which is described in the great Indian poem, the Mahabarata. The period of Yudhisthira is here stated to be 2526, meaning from the beginning of the Kaly Yuga, the epoch of which is 3102 B.C., which would make the date of Yudhisthira and Parasara 576 B.C.*

But there are two circumstances in the above statement by Varaha which bear a suspicious character, and which may have led Colebrooke to hesitate about receiving it as an authenticated fact. Varaha states (1) that Yudhisthira was the ruler when the Munis or Rishis were in Magha; he also says (2) they remain for a hundred years in each Asterism. Now, if Varaha intended by this that the Munis, which are fixed stars, were moving through the fixed Asterism Magha, no intelligible meaning could be attached to the statement; but if, as before explained, it was the Solstitial Colure to which he referred, and which may have got the name of the moveable line of the Munis when it coincided with the fixed line of the Rishis, in about 1590 B.C. or 1630 B.C., the Vernal Equinox would

---

* In the Ayeen Akbery, II., p. 110, it is stated that the great war "happened in the end of the Dwapar Yuga, 105 years prior to the commencement of the Kali Yuga, being 4831 years anterior to the fortieth year of the present reign" (that of Akber).

The fortieth year of Akber was 1595 A.D.

∴ 4831 − 1595 A.D. = 3236 B.C. − 105 = 3131, commencement of the Kali Yuga.

The commencement of the Kali Yuga being reckoned to be 3131 B.C.
Era of Yudhisthira, according to Garga, per Vahara Mihira, 2526 „

Therefore .. .. 605 „

have gone back through 30° from 3° 20′ of the Lunar Asterism Crittica to the first point of Aswini, and the Solstice in the same time would have retrograded through an equal arc of 30° from the first of Magha to 10° of Punarvasu.

The received opinion, however, as before stated, is that Yudhisthira (with Garga and Parasara) lived some time about the 12th or 13th centuries before the Christian Era, whilst Davis believed the date of Parasara to be even as early as 1391 B.C.

Contemporary with Parasara, the epoch, also, of the Indian Prince Purasurama is supposed to have been established.

Purasurama is described as a great encourager of astronomy, and is said to have lived about 200 years before Rama.

Dr. Buchanan, in his "Journey to Malayala" (September, 1800) states that the astronomers there reckoned by cycles of 1,000 years from Purasurama, and that of the then current cycle, 976 years were expired in September, 1800, and that 2,976 must have elapsed from the epoch of Purasurama to the year 1800 A.D.; from which it is concluded that the epoch of that prince is 1176 B.C.

The years of this epoch of Purasurama are reckoned as beginning with the sign of Virgo, or, rather, with the month of Aswina.

According to Hindu tradition, Purasurama was a Brahmin, who had a great contest with the Kshetrias, whom he vanquished, and he also reduced to subjection the Sanchalas, a wild and barbarous nation said to feed on human flesh.

One of the many Grotto temples of Ellora appears, from an illustration by Daniel, to go by his name, from which it may seem that he was of an earlier date than that when the temples were excavated; but not of a date so old as the grotto temples of Elephanta, or the still older mixture of large Buddhist sculptures in the grotto temples of Salsette and Karli, immense undertakings which furnish, like the pyramids, convincing proof of the long duration of time required for their construction. Whatever conjectures may be made concerning Purasurama and his time, no one doubts that at some early period of Hindu history, perhaps more definable than in the case of men who have been mentioned, the famous Rama, the hero of so many adventures, as related in the Ramayana and other poetical works of the Hindus, had a real existence as a sovereign ruler of Ayodhya; but the period when he lived is a question, nevertheless, not easily determined.

The Hindus place him at some time between the Indian silver and brazen ages. Sir W. Jones, in a table of his chronology, gives, as the best of two supposed approximate dates, the year 1399 B.C.

During the time of Rama, and that of his father, Dasaratha, a wise and pious prince, the study of astronomy received much encouragement, and was cultivated with much attention. It has been seen that Sir W. Jones could derive no authentic information regarding him from Hindu chronological records, and, indeed, it is improbable that such information could be derived from a source which is mostly of a fictitious character. But if the Lagna or Horoscope of Rama be correctly recorded in the Ramayana, there can be no difficulty in fixing the date of his birth.

Bentley, from such a source, calculated that Rama was born on the 6th April, 961 B.C.

This date can be easily verified by reference to the position which each of the celestial bodies—the sun, the moon, and the five planets —were stated to have occupied on the ninth lunar day of Chaitra, the sun being then in Aries, the moon in Cancer, Venus in Pisces, Jupiter in Cancer, Mars in Capricorn, and Saturn in Libra. Bentley suspects that these positions were obtained as the result of modern calculation, and not by actual observation, for he "thinks that the signs were not known by these names in the time of Rama. But, whether from computation or otherwise, they point out that Rama was born on the date which he has given."

Further:—"When Rama attained the age of manhood his father, Dasaratha, in consequence of certain positions of the planets, approaching to a conjunction, supposed to portend evil, wished to share the government with him, Dasaratha says: 'My star, O Rama, is crowded with portentous planets—the sun, the moon's ascending node, and Mars. To-day the moon rose in Punarvasu, the astronomers announce her entering Pushya to-morrow; be thou installed in Pushya. The sun's ingress into Pushya being now come, the Lagna of Karkata (the sign Cancer, in which Rama was born) having begun to ascend above the horizon, the moon forbore to shine; the sun disappeared, while it was day, a cloud of locusts, Mars, Jupiter, and the other planets inauspicious approaching.'"

The facts pointed out here (says Bentley) show that there was an eclipse of the sun at or near the beginning of Karkata at the moon's ascending node (Rahu being present), and that the planets were not far distant from each other. From these circumstances, he calculated the time to have been the 2nd July, 940 B.C., and that then Rama was one and twenty years old.

Here, however, Bentley points out that the beginning of Pushya and that of Carcata, or Cancer, were supposed to coincide. This implies that the Summer Solstice was then the first of Pushya, which would make the position of the Equinoctial point then only 3° 20′ short of the first of Aswini. The time for a regression through 3° 20′ has been shown before to be about 240 years, which, taken from the epoch 570 A.D., leaves 330 A.D. as the time when the first of Carcata coincided with the first of Pushya.

Bentley makes the date 295 A.D. (upwards of 1,000 years after the time when the recorded deeds are supposed to have happened), and from it he infers that this was the period when the Ramayana was written. And he says:—"In giving the age of the Ramayana of Valmika, as it is called, I do not mean to say that the facts on which that romance was founded, in part, did not exist long before. On the contrary, my opinion is that they did, and probably were to be found in histories or oral traditions brought down to the time. The author of the Ramayana was more a poet than an astronomer, and, being unacquainted with the precession, he fell into the mistake alluded to, for I do not suppose it was intentional, as that could answer no purpose."

There is another circumstance which (Bentley says) must have occurred in the time of Rama, *i.e.*, the fiction of the "Churning of the Ocean," founded upon the various incidents of an eclipse of the sun, which took place, according to his calculation, when the Vernal Equinox was in the middle of the Asterism Bharani, in the year 945 B.C., on the 25th October.

It is a highly-coloured fable (an allegory of an eclipse) in poetical language—a pretended fight between the Suras and the Asuras, the Gods of Light and Darkness, and their offspring.

An account of it is given in the Puranas, but it is more fully described in the great poem of the Hindus, the Mahabharata (B. 1 chap. 5), a translation of which is given by Wilkins, and transcribed in full by Bentley.

In this eclipse Saturn was discovered. He is said to have been "born in the moon's shadow, which pointed towards the Lunar Asterism Rohini."

The name given by the Hindus to the new planet was Chaya-Suta (Offspring of the Shadow).

It is further supposed as probable that the Theogony of the Hindus was invented at about this period (945 B.C.), and that the heavens were then divided, and shares of it assigned to the several Gods.

A translation from Hesiod, given by Bentley, describes the war between the Gods and the Giants, a fiction resembling that concerning the "Churning of the Ocean" of the Puranas, the former being supposed by Bentley to be borrowed from the Hindu fable at a period some 200 years later, or about 746 B.C. Bentley, in this connection, enters largely into a comparison of the mythology of the Hindus, the Chaldeans, the Egyptians, and the Greeks; but it will be needless in this work to follow him in such a comparison.

It is also stated by Bentley that in the same year (945 B.C.), according to observations then recorded, the Solstitial Colure cut the Lunar Asterisms Aslesha in 3° 20′ and Sravana in 10°.

Reckoning from the fact that the Hindu solar months always begin at the moment the sun enters a sign of the Zodiac, and the day on which the eclipse happened being the 23rd of the month Kartica, it is deduced that the first day of the month fell on the sixth day of the moon. "This being the time of the Autumnal Equinox, it was found by observation that the Colures had fallen back in respect of the fixed stars 3° 20′ since the former observations in 1192 B.C."

It will be observed that, from this retrograde motion of 3° 20′ in 247 years, the mean annual rate of precession (48[illegible]″) may be readily found. "In the same period of 247 years and one month they found that the moon had made 3,303 revolutions, and one sign over, that there were also 3,056 lunations or synodic periods, and the number of days in the whole period was 90,245¼."

From these data it is easy to deduce:—

		Days.	hrs.	mins.	secs.
The length of the tropical year ..	—	365	5	50	10[illegible]
" " " sidereal " ..	—	365	6	9	52[illegible]
" moon's tropical revolution ..	—	27	7	43	5
" lunar month	—	29	12	44	3

Now, there is nothing improbable regarding the observations thus stated by Bentley. They are just the kind of observations the Hindu astronomers were constantly making, to determine the days when the sun was in an Equinox or a Solstice—those four days of the year when sacrifices and offerings were to be made to the Supreme Being—observations which, as expressed by Laplace, resulted in "the remarkable exactness of the mean motions which they (the Hindus) have assigned to the sun and the moon, and necessarily required very ancient observations."

Bentley, moreover, from a study of the ancient Hindu calendars, and from the circumstance that the period of 247$\frac{1}{12}$ years contained a month "more than 247 years, considered it obvious that this period must begin and end with the same month of the year and that the next succeeding period would begin with the month following, and thereby change the commencement of the year one month later each period; and, moreover, as there was a complete number of lunations (3,056) in the period, it follows that the moon's age would be always the same, at the commencement of each succeeding period."

This would prove that in these early times the solar year was tropical, and estimated from Equinox to Equinox, just as it is in our modern system, only that its beginning would be a month later in each period of $247\frac{1}{12}$ years.

In accordance with this statement he gives the following "Table of all the changes made in the commencement of the Hindu year from 1192 B.C. down to 538 A.D., when the ancient method was entirely laid aside, and the present, or sidereal astronomy introduced:—

Periods.	Began.	Months beginning the Period.	Lunar Asterism coinciding at the beginning.	Sun's longitude at the beginning.	Moon's longitude at the beginning.	Calendar.	Corresponding day of European months.
1	1192 B.C.	Aswina ..	Chitra	5s 0°	7s 12°	Shasty Adikalpa ..	1st Sept.
2	945 ,,	Kartika ..	Visakha.. ..	6 0	8 12	Guha Shasti	1st Oct.
3	698 ,,	Agrahayana	Jyestha	7 0	9 12	*Mitra Saptami ..	29th Oct.
4	451 ,,	Pausha ..	P. Ashadha ..	8 0	10 12	....	27th Nov.
5	204 ,,	Magha ..	Sravana	9 0	11 12	*Bhascara Saptami	25th Dec.
6	44 A.D.	Phalguna ..	Satabisha ..	10 0	0 12	....	23rd Jan.
7	291 ,,	Chaitra ..	U. Bhadrapada	11 0	1 12	....	21st Feb.
8	538 ,,	Vaisakha ..	Aswini	0 0	2 12	*Jahnu Saptami ..	22nd Mar.

The preceding table, founded upon eight successive periods, each of $247\frac{1}{12}$ years, derived from corresponding retrograde progressive move-

* Names of the Sun.

ments of the Equinoctial and Solstitial Colures, forms the basis of Bentley's system of astronomical chronology of the Hindus, terminating with the date 22nd March, 538 A.D.

The Hindus themselves state that the great change in their astronomical system, from a moveable to a fixed origin, a point of the Ecliptic, from which their longitudes are now reckoned, was made when the Vernal Equinox coincided with the first point of the Lunar Asterism Aswini. The date when this occurred is stated as above, by Bentley, to have been 538 A.D.; by Colebrooke, from the mean of two calculations, 582 A.D.; by the American translators of the Surya Siddhanta, 570 A.D., which is reckoned from the longitude of ζ Piscium.*

The differences in these several dates arise principally from the different estimates of the precession which are used in the respective calculations, for a variation of half a second in the precession produces a difference of twelve years in the calculation over 1,200 years

* The longitude of ζ Piscium in 1800 was about 17° 6′, it is of the fifth magnitude, and 110′ West of the beginning of Aswini.

CHAPTER X.

RISE OF THE BUDDHIST HERESY AND ITS EFFECT ON HINDU ASTRONOMY.

[*B.C.* 945—200.]

According to Fergusson, "the inhabitants of the Valley of the Ganges, before the Aryans reached India, seem to have been tree and serpent worshippers, a people without any distinct idea of God, but apparently worshipping their ancestors and, it may be, indulging in human sacrifices."

Undoubtedly, however, when the Aryans spread into the country, as we have seen, from the North-west, there arose the religion of the Brahmins as the dominant faith extant in the periods referred to in the last chapter. The evidences afforded by the contents of the Vedas and the Institutes of Menu, which are almost universally regarded as having been compiled prior to 1000 B.C., incontestibly prove the then existence of the Brahminical faith as an organised and settled system, although probably much of the antecedent savage worship still remained.

No reference, however, can be found in the Hindu writings of later date (so far as the writer has been able to ascertain) of any authentic value, to the period which succeeded that (circ. B.C. 945), dealt with in the Ramayana, until the appearance of Buddha.

"In the sixth century B.C.," continues Fergussion, "Sakya Muni (Buddha) reformed this barbarous fetishism into a religion now known as Buddhism, and raised the oppressed inhabitants of Northern India to the first rank in their own country. . . . . The castes of the Aryans were abolished. All men were equal, and all could obtain beatitude by the negation of enjoyment and the practice of prescribed ascetic duties."

Buddhism, as introduced by Sakya Muni, appears to have spread into the North-east of India, Cashmere, Thibet, Burmah, and to

China. The remains, still existing, of Buddhist temples, and the names to be found in the several districts, appear also to indicate that this religion extended across Central India to the Mahratta country, to Malwa, the Deccan, and to states bordering on the Nerbudda River (a name of some significance in this connection), and further, to Western India, and finally to Ceylon. In this progress, its votaries established themselves at various centres, such as Dhar, Baug, Ellora, Bhilsa, and, in some measure, at Oojein, each of which places has its own fragmentary history, separately from the others, in connection with the rise and establishment of Buddhism in the country.

We have, however, no authentic history of India previous to the invasion of Alexander in 350 B.C.

From the officers and men of science, who accompanied him in his expedition, we gather some information of much value regarding the condition of the people at this time. The account represents them as a great and powerful nation, the country as divided into a number of kingdoms of great extent, and population—a description which implies that it must have taken long periods of time for the growth and consolidation of the then nation. The Greeks, also, accumulated much information regarding the physical character of the territories which they had entered, with respect to soil, productions, and climate, though the extent of country over which they had an opportunity of forming an opinion was limited to a portion of the Punjab, and to the borders of the provinces through which they passed, along the banks of the Indus, in the famous voyage down 1,000 miles of that river, which took them nine months to reach the ocean.

In the brief time during which they remained in the country they learned that the people were divided into four classes, or castes.

The highest, as a sacred body of divines, held supremacy over the rest. It was their province to study the principles of religion, to conduct its offices, and to cultivate the sciences, in the capacity of priests, philosophers, and teachers.

The class next in order, the warrior caste, held the position of rulers and magistrates in times of peace, and of commanders and soldiers in war.

The third class consisted of the husbandmen and merchants, and the fourth of artisans, labourers, and servants.

They noted also the character of the inhabitants, their political and social institutions, their manners and customs, with everything else that came under their own particular observation.

These were all narrated with minute accuracy, insomuch that, to our countrymen who have been long familiar with similar things, it has appeared wonderful how little they are changed from what they were twenty-two centuries ago.

Soon after the death of Alexander, the several kingdoms which had opposed him became united under one ruler, a man of low origin, who had usurped the throne of Maghada, after killing his own sovereign. This monarch, called by the Greeks Sandracottus, but named in India Chandra Gupta, became king in 343 or 315 B.C. His court was held at Pataliputra, or Palibothra, a city described as being exceedingly large and populous, whose site is now unknown, but believed to be that on which Patna is situated.

Both Chandra Gupta and his son, Bindusara, appear to have been Hindus of the true orthodox faith; but Asoka, the grandson of Chandra Gupta, became a convert to Buddhism, between the partisans of which persuasion and those of the Brahminical faith a long controversy had existed.

On the death of his father (circ. 266, or 263 B.C.), Asoka became the great patron of the new faith, and presided over the third Buddhist* Council held in 245 or 242 B.C.

* The founder of this religion, Sakya-Muni, or Gautama Buddh, was a prince of the Solar dynasty, "who for a long period subsequent to the advent of the Aryans into India, had held permanent sway in Ayodia—

Professor Max Muller observes:—"Though Buddhism became recognised as a state religion, through Asoka, in the third century only, there can be little doubt that it had been growing in the minds of the people for several generations, and though there is some doubt as to the exact date of Buddha's death, his traditional era begins 543 B.C., and we may safely assign the origin of Buddhism to about 500 B.C."

Bentley says, regarding the fifth astronomical period of his Chronology, marked in his table (p. 124) as beginning on the 25th December, 204 B.C., that the "Hindu year began with the month of Magha, at the Winter Solstice, and in the first point of the Lunar Asterism Sravana, marked in the calendar with the word Makari Saptami, denoting that the sun entered Capricorn on the seventh of the moon. Sometimes it is marked Bhaskara Saptami."

Now, since the third Buddhist Council is stated by Max Muller to have been held in 245 or 242 B.C., and by Fergusson at about 250 B.C. (at which period Buddhism was the state religion), and from early caves of Behar, Fergusson deduces that the time of Dasaratha, the grandson of Asoka, must have been about 200 B.C.,

---

the modern Oude. About the 10th or 12th century B.C., they were superseded by another race of much less purely Aryan blood, known as the Lunar race, who transferred the seat of power to capitals situated in the northern parts of the Doab. In consequence of this the lineal descendants of the Solar kings were reduced to a petty principalty at the foot of the Himalayas, where Sakya-Muni was born about 623 B.C. He spent many years in meditation and mortification as an Ascetic, to fit himself for the task of alleviating the misery incident to human existence, by which he had become painfully impressed, and for forty-five years he steadily devoted himself to the task he had set before himself, wandering from city to city, teaching and preaching and doing everything that gentle means could effect to disseminate the doctrines which he believed were to regenerate the world and take out the sting of human misery."

The date of his death has been estimated both by General Cunningham and Professor Max Müller to have been 477 B.C., but no certainty is entertained as to the period.

x

it follows that Dasaratha must have lived at a period somewhere near the beginning of Bentley's fifth astronomical period.*

With regard to the position of Indian literature at this time, Professor Max Muller says, "that in the third century B.C., the ancient Sanscrit language had dwindled down to a mere Volgare, or Pracrit, and the ancient religion of the Veda had developed into Buddhism, and had been superseded by its own offspring, the state religion of Asoka, the grandson of Kandra Gupta."

The subjects of discussion in the controversy between the Orthodox Brahmins and the Buddhists appear to have been principally on the assumed divine authority of the Vedas, the utility of sacrifices and ceremonies for the dead, and on the iniquity of killing animals for food. The Buddhists, moreover, were regarded as heretics by partisans of the other faith.

---

* Ferguson, a great authority on the Architecture of India, remarking upon the times here referred to, says:—"The Aryans wrote books but they built no buildings. Their remains are to be found in the Vedas and the Laws of Menu, and in the influence of their superior power on the lower races; but they excavated no caves, and they reared no monuments of stone or brick that were calculated to endure after having served their original and ephemeral purpose.

"Our history (Indian Architecture) commences with the Architecture of the Buddhists. Some of their monuments can be dated with certainty as far back as 250 B.C., and we not only know from history that they are the oldest, but they bear on their face the proofs of their primogeniture. Though most of them are carved in the hardest granite, every form and every detail is so essentially wooden, that we feel in examining them that we are assisting at the birth of a new style.

"The circumstances of the architectural history of India, commencing with Asoka, about 250 B.C., and of all the monuments for at least 500 years after that time being Buddhist, are two cardinal facts that cannot be too strongly insisted upon or too often repeated by those who wish to clear away a great deal which has hitherto tended to render the subject unintelligible.

"The principal monuments by which Asoka is known to us are his inscriptions. Three of these are engraved on the living rock, one near Cuttack, on the shores on the Bay of Bengal; another near Ionaghur, in

About 204 B.C. (according to Bentley) "improvements were made in astronomy; new and more accurate tables of the planetary motions and positions were formed, and equations introduced. Besides these improvements, the Hindu history was divided into periods for chronological purposes, which periods, in order that they might never be lost, or, if lost or disputed, might, with the assistance of a few data, be again recovered, were settled and fixed by astronomical computations," in the following manner:—

"The years with which each period was to commence and end, having been previously fixed on, the inventors then, by computation, determine the month and the moon's age, on the very day on which Jupiter is found to be in conjunction with the sun in each of the years so fixed on;" which, being recorded in the calendar and other books, might at any time be referred to for clearing up any doubt, in case of necessity.

For two or three centuries before Asoka began his reign, there is an unaccountable dearth of information regarding the astronomy of that period. Bentley suspected that there had been a great destruction of manuscripts.

He states that there is still a tradition that the "Maharastras, or Maharattas, destroyed all the ancient works, that the people hid their books in wells, tanks, and other places, but to no purpose, for

---

Guzerat, 1,000 miles of the last; and a third at Kapur di Giri, 900 miles north of Ionaghur.

"Slightly more architectural than these are the Lâts or pillars, erected to contain edicts conveying the principal doctrines of the Buddhist religion as then understood.

"One of them is at Delhi, having been re-erected by Feroze Shah in his palace, as a monument of his victory over the Hindus.

"Three more are standing near the River Gunduck, in Tirhoot; and one placed recently in the Fort of Allahabad. A fragment of another was discovered near Delhi, and part of a seventh was used as a roller on the Benares road by a Company's officer."—*Ferguson's Indian Architecture, Vol. II., p.* 458.

K 2

hardly any escaped, and those that did then escape were afterwards picked up by degrees, that none were allowed to be in circulation. . . . Which will (he says) account for the paucity of ancient facts and observations that have reached our times."

In the above period he could find only one observation worth mentioning, in 215 B.C., when it was found that, at the Winter Solstice, or the beginning of the solar month Magha, the sun and moon were in conjunction at sunrise on a Sunday.

It may be as Bentley has remarked, that there was a search made for manuscripts at this period of Hindu astronomy; but certainly it would not be made for the purpose of destroying them, and it is probably owing to such a search that so many manuscripts have been preserved.* It may be that the search was made at the instance of the learned men of the time, for the purpose of restoring their ancient literature and science. Amongst the Jyotishticas, the manuscript relics would be preserved with care, but many of the families would have become extinct, and their writings would have found their way into foreign hands. Many, also, would have been lost. Nevertheless, there can be no doubt that in the ancient writings of that period were found the materials from which were compiled and condensed the relatively correct mean motions of the planets, and the rules of astronomy and mathematics, given in the

---

* The number of separate works in Sanscrit, of which manuscripts are still in existence, is estimated by Professor Max Muller to amount to about 10,000, which makes him exclaim, "What would Plato and Aristotle have said, if they had been told that at their time there existed in that India which Alexander had just discovered, if not conquered, an ancient literature far richer than anything they possessed at that time in Greece?"

We can readily conceive that amongst these manuscripts there are dramas and works of fiction innumerable, and treatises on literature and science, but there is little hope of their being completely investigated and sifted, and only like nuggets in a mine are the really valuable works likely to be found accidentally.

text books preserved in different parts of India, and from which were computed the various tables already mentioned. Some of these were carried into Siam (probably by ancestors of the Buddhist priesthood of that country, where they were finally driven from India), and bear evidence of their Indian origin, by the corrections required for the difference of longitude of places, in the two countries. It is not improbable, also, that the Buddhists who found a home in China when their rulers were compelled to retire from Hindostan, in the persecution instigated by Sancara, and Udayana Acharya, by princes of the Vaishnava and Saiva Sects, carried with them tables of a like character. Again, there were, no doubt, among the manuscripts sought, many ancient mathematical works which have been cited as authorities in later works, and which must have been in existence at the times when they were quoted in such later works.

Thus, in 1150 A.D. we find Bhaskara mentioning the names of a number of works on algebra, which he must have had in his possession when he wrote the following lines at the conclusion of his work on the Vija-Ganita, or Algebra, commending his elementary work:—

(218) "As the treatises of Brahmegupta, Sridhara, and Padmanabha are too diffusive, I (Bhaskara) have compressed the substance of them, in a well-seasoned compendium, for the gratification of learners."

(219) "For the volume contains a thousand lines, including precept and example. Sometimes exemplified, to explain the sense and bearing of a rule, sometimes to exemplify its scope and adaptation; one while to show variety of inferences; another while to manifest the principle. For these, there are no end of instances, and therefore a few only are exhibited."

In the body of the Vija-Ganita, Bhaskara also cites from Sridara and from Padmanabha's Algebra, and repeatedly refers to other writers in general terms. Where his commentators, who must have had the works in their possession, understand him to allude to Aryabhatta, Brahmegupta, and the Scholiasts, Chaturveda, Prithudaca Swami, and other writers, Colebrooke says:—"A long and diligent research in various parts of India has, however, failed of recovering any part of the Padmanabha Vija, or Algebra, of Padmanabha, and of the algebraic and other works of Aryabhatta."

There can, however, be no doubt that, judging from the extracts given, and from references of Bhaskara and other writers, many previous works were existing, and some of them doubtless compiled at the time (200 B.C.), when, as Bentley says, there was a great revival and reconstruction of the Hindu astronomy of that date.

Colebrooke was more fortunate in regard to Sridhara and Brahmegupta. He possessed Sridhara's Arithmetic and an incomplete copy of the text and Scholia of the Brahma Siddhanta, revised and edited by Brahmegupta, which will be more particularly described further on.

In the third volume of the "Asiatic Researches," Davis, near the end of the last century, referring to the ancient writers named by Colebrooke, remarks that "almost any trouble and expense would be compensated by the possession of the three treatises on algebra from which Bhaskara declares he extracted his Vija-Ganita, and which (in Bengal) are supposed to be entirely lost."

The suggestions that enquiries should be made for the works by Europeans who had access to Oojein have been acted upon, but without success.

Thus, the paucity of material supplied to the narrator of the events relating to the period under consideration, forbids an exhaustive discussion of the state of astronomy at that date; but much may, nevertheless, be gathered from isolated circumstances.

The success of the revolution which made Buddhism the state religion is supposed to have been owing to a great increase in the population, and a wide-spread discontent amongst the lower orders, which found in the new government a relief from the severe discipline of the Brahminical and other higher classes. Caste was abolished, and the freedom of the subject asserted. The overthrow of the Aryan rule was, therefore, easily accomplished. We can only form conjectures regarding the attitude of the Brahmins in this conjuncture of their affairs. It may be that some of them temporised with the ruling powers, compelled by circumstances to conform to the spirit of the times, and appeared as converts to the new faith, concealing their opinions to avoid persecution. Others appear to have turned their attention to literary pursuits, and created those allegories, fables, and tales of fiction, which have since been the amusement and formed the mythology of the country for many centuries.

It does not, however, appear that cruelty or persecution was ever practised towards their adversaries by the Buddhists. The tendency of their religion was to promote the advancement of their faith by gentle means, and to obtain proselytes by persuasion. The Brahmins, therefore, appear to have been allowed to carry on their studies in peace and without molestation.

Their Sanscrit schools must have been conducted upon nearly the same principles as they are now, the love of their ancient language, descending in families which traced their lineage backward to men who have been distinguished for their learning at various periods of their history; and at all times there were amongst them men learned in the four Vedas, and who had attained to the rank and title of Acharya.

It is surmised that at some time during the Bhuddist supremacy the various sects of astronomers of that period were led by their liberal rulers to a freer intercourse with each other, that the toleration which they themselves appear to have practised was encouraged among the teachers of different systems of astronomy, who were allowed to discuss their diverse doctrines, or to discourse upon them at the courts of the Buddhist princes. This tended to promote a better understanding of their respective systems, from which a mutual improvement in methods of calculation and observation was adopted in their teaching and in their text books.

It was by means of these Sanscrit seminaries (which have existed from time immemorial) that their learning has been transmitted from age to age, either orally or by writing, and condensed to the merest formulæ in words, seemingly framed for the memory, as a supplement to the teachers' lectures, and which would supply the explanations and proofs necessary for fully understanding their import and uses.

In the Buddhist period the men of learning, sages among the Brahmins, retired with their disciples and adherents to their quiet rural homes, and there composed, compiled, or revised, many of the works of literature and science. These have escaped the ravages of time and the many vicissitudes to which they have been exposed during so many subsequent centuries.

From an examination of these works, it would appear that their teaching in astronomy was theoretical, founded upon rules which had been constructed by the Munis of preceding ages, and their calculations were based on astronomical tables that had become obsolete, consequent upon a severance of theory from practice. Through neglect in applying the necessary corrections (Bija) to the mean motions, errors in them, multiplying and increasing year by year, would seem to have caused the whole system of their astronomy to become so confused, that rules relating to conjunctions, oppositions, and other phenomena, though demonstrable and true (when applied with correct numerical constants), became now completely at variance with the facts of observations, and hence the calendar became utterly untrustworthy.

In the subsequent reconstruction of their astronomy, there is evidence of the great use that was made of the Vija-Ganita, the ancient algebra of the Hindus, regarding which the testimony in every form tends to show that it had its origin in India; and to the ancient astronomical works were often appended separate treatises, in the form of chapters on arithmetic, algebra, mensuration, spherics and trigonometry.

We are informed by Indian authorities that the earliest known uninspired writer on astronomy was a mathematician named Aryabhatta, who lived near about this Buddhist period. He seems to have been of a sect somewhat different from the orthodox Brahminical astronomers of a later time. He was skilled in the application of algebra to questions in astronomy, and is reported to have been acquainted with an analytical method of solving problems, which went by the name of Cuttaca (translated, "Pulveriser," from Cutt, to grind, or Pulverise)—a method resembling one of our own for the solution, in all cases, of indeterminate equations of the first degree. This method will be described more fully hereafter in connection with the Brahma Siddhanta.

Aryabhatta wrote a number of works on astronomy, which are now known only by quotations from his writings, given by Brahmegupta and other subsequent astronomers, for the purpose of controverting the doctrines maintained in them. It is in general by these citations that Aryabhatta was known as a very eminent astronomer, who was at least anterior to Brahmegupta, and probably flourished in the beginning of the Christian Era.

Colebrooke, from various considerations, concluded that he must unquestionably be placed "earlier than the fifth century of the Saca, and probably so, by several (by more than two or three) centuries, and, not unlikely, before either Saca or Sambat eras.*

"In other words, he flourished some ages before the sixth century of the Christian Era; and perhaps lived before, or at latest soon after, its commencement."

From the quotations of Brahmegupta we learn that Aryabhatta maintained the diurnal rotation of the earth round its axis. The starry sphere "he affirms is stationary, and the earth, making a

* Sambat era, 56 B.C., Saca era, 78—9 A.D.

revolution, produces the daily rising and setting of the stars and planets."

To which Brahmegupta answers:—"If the earth move a minute in a prana, then whence and in what route does it proceed? If it roll, then why do not lofty objects fall?"

The commentator of Brahmegupta, Prithudaca Swami, replies:—"Aryabhatta's opinion appears, nevertheless, satisfactory, since planets cannot have two motions at once; and the objection that lofty things would fall is contradicted, for every day the under part of the earth is also the upper, since, wherever the spectator stands on the earth's surface, even that spot is the uppermost point."*

From numerous quotations of Bhattopala and other eminent Indian mathematicians, it was also known that "Aryabhatta accounted for the diurnal rotation of the earth on its axis, by a wind or current of aerial fluid, the extent of which, according to the orbit assigned to it by him, was little more than one hundred miles from the surface of the earth."

Also that he possessed the true theory of the causes of "lunar and solar eclipses, affirming the moon to be essentially dark and only illumined by the sun; that he noticed the motion of the Solstitial and Equinoctial Points, but restricted it to a regular oscillation, of which he assigned the limit and the period; that he ascribed to the Epicycles, by which the motion of a planet is repre-

---

* The theory that the earth moves daily round an axis, and that it has a motion round the Sun as a kind of centre, which is completed in a year, is a doctrine so far removed from the evidence of our senses and so contrary to our daily observations, that before the proofs are understood, if it is received at all, it will be received as a mere opinion of men better able to judge of such matters, which may or not be true.

In the times of Copernicus, Kepler, and Galileo, it is not therefore astonishing that a theory which was so contrary and so entirely opposed to that which had been so universally received, and which had prevailed in all countries from the beginning of the world, should have been met with ridicule, and with the persecution of its authors and their followers.

sented, as a form varying from the circle and nearly elliptic; that he recognised a motion of all the nodes and apsides of all the primary planets, as well as of the moon.

"The text of Aryabhatta specifies the earth's diameter 1,050 Yojanas, and the earth's orbit, or circumference of the earth's wind, 3,393 Yojanas.

"The ratio here employed of the circumference to the diameter is 22 to 7, an approximation nearer than that which both Brahmegupta and Sridhara employ in their mensuration.

"He treated of algebra, etc., under distinct heads of Cuttaca, a problem for the resolution of indeterminate ones, and Vija, principles of computation, or analysis in general."

Aryabhatta is cited, according to the statement of Colebrooke, by a thousand Hindu writers on astronomy, "as the author of a system and founder of a sect in this science."

It is stated that about the middle of the eighth century of our era the Arabs first became acquainted with the astronomy of the Hindus. A Hindu astrologer and mathematician was drawn to the court of the Abbasside Khalifs, Almansur and Almamun, and, by order of the Khalifs, the Indian mean motions of the planets were made the foundation of the Arabian astronomical tables.

At this time the difficulty of obtaining an insight into the Indian sciences was made the subject of complaint by the Arabic authors of the Tarikhul hucama, who assigned as the cause the distance of the countries and the various impediments to intercourse.

The three primitive sects into which the Indian astronomers were divided differed from each other, amongst other things, in their mode of beginning the astronomical day, and their names were Audayaca from Udaya rising. Ardharatrica from Ardharatri Midnight, and Madhyandinas from Madyandina Mid-day.

The founder of the first of these sects was Aryabhatta, who is said to have had more correct notions of the planetary motions than any of the writers who lived in later ages. He is mentioned as having made corrections in a system received by him from earlier sources, and referred to as that of Parasara, from which he took the numbers for the mean motions of the planets. It is probable that the earlier sources here referred to were his father and other astronomers, who were living during the period of the Buddhist supremacy or afterwards, when Vicramaditya became the ruler at Ujein.

It is probable that about this time (200 B.C.), when the revival of the Hindu astronomy began, the allegory of the death of Durga was invented by the Brahmins for the purpose of keeping in remembrance the decadence of their favourite science, and its subsequent revival.

The death of Durga is still sometimes represented in private spectacles, wherein large figures are constructed to take part in tableaux illustrating some of the scenes described in the Ramayana and the Mahabharata, such as Rama's lament over the death of Luksmi, and others of a like nature. Plate XII. is taken from a photograph of figures representing the calamity which overtook Hindu astronomy at this eventful period.

The great importance given to time as a mighty worker of events was well understood in its personification as Siva. Years were personified as his wives, one of whom, Kalee, was described as an insatiable monster devastating whole countries, which was in earlier times but a figurative way of expressing that such and such years had been calamitous in famines, pestilence, and wars, which would have depopulated the world, had not Brahma personally interfered with Siva and induced him to keep his wife in order. Siva, bewildered, had no other means of stopping her madness than by throwing himself at her feet, and only as she was stepping on his body did she become aware of the disrespect she was showing to her husband; and, from shame, she then ceased from her devastations.

Durga, an astronomical representation of the year, and also a wife of Siva, was of a higher caste. She was the daughter of

PLATE XII.

DEATH OF DURGO.

Daksha (a representation of the Ecliptic). But Siva was regarded by the higher class of gods as a dissolute character, with snakes and other reptiles crawling over him, alluding to his worshippers, the Nagas, who were devoted to his service.

One tradition regarding Siva was that his father-in-law, Daksha, had invited to a great feast all the gods, celestial and terrestrial, the planets, stars, Rishis, and Munis, with their worshippers. The feast, as a figure, was intended to show the importance attached to astronomy, but without reference to time, which was an insult to Siva. This gave great pain to Durga, who, after much entreaty, was permitted to present herself at her father's house, and to appear in the assembly; but such was her distress at witnessing the contempt shown towards her husband that she died of grief. In other words, the year (which, in the ancient astronomy, had been derived from the Ecliptic, by means of a long series of observations on the sun, moon and stars, and had become so exact in length that predictions and calculations having reference to the times of the year could be depended upon to agree with the events), had been lost. Through disregarding the effects produced by time, and neglecting to apply the necessary corrections to their calculations, so many errors had crept into the predictions of the calendar that even the length of the year itself became unknown—or Durga died.

To revenge her death, Siva, from his own body, created a numerous army, by means of which all the gods who had assembled at the feast of Daksha were destroyed.

The meaning of this is that a multiplicity of errors arose in computations regarding the planets, seasons, and months, causing the greatest confusion in periods of religious observances, until at length no regard was paid to astronomical observations, and all knowledge of the celestial sphere of the Ecliptic, and of the planetary motions was lost. Astronomy was no longer studied. Daksha, with all the other Celestial Deities, were slain.

At length, Brahma, moved with compassion, caused Siva to relent.

A search was made for the bodies of the dead; a restoration of nearly all the gods to life was effected; but when it came to the turn of Daksha, his body was found without a head. A goat was, however, found near; its head was cut off, and Daksha was restored to life with the head of a goat. He retired to Benares, where he has often been seen since, wandering about with his goat's head, and looking very sheepish.

This part of the legend, no doubt, alludes to the revival of the study of astronomy, and is intended to show that difficulties had arisen between the astronomers of Ojjain and those of Benares regarding the beginning of the Ecliptic. The question was whether it should begin with the Equinox at Aries, or with the Solstice of Capricorn. The sect which made use of the Solar Zodiac would adopt the latter, and would have placed the beginning (or origin of longitudes) on the system of Lunar Asterisms at 3° 20′ of the Asterism Uttarashadha, which was by no means a suitable origin for this lunar system. It seems to have been, therefore, finally decided, at Oojein at least, that the years and the Ecliptic should both commence at Aswini, which made also this beginning coincide with the first of Mesha (Aries), then also the Vernal Equinox; although from the legend it might appear that some of the astronomers of Benares still held the beginning of the year to be at Capricorn.

From this difference of opinion we might almost infer that the system of Lunar Asterisms was that of the Brahmins of Arya-Verta, and that the system which had for its foundation the Solar Zodiac was that used by a sect of astronomers of the more Northern parts of India; and it may have been this difference of their systems which originated the distinction between the solar and lunar races of India.

Aryabhatta may have been one of the Northern sect, for he held what were considered to be unorthodox opinions in astronomy, which were cited by Brahmegupta, not for the purpose of praising or approving of them, but for contesting and controverting them.

CHAPTER XI.

PERIOD FROM THE RESTORATION OF THE POWER OF THE BRAHMINS TO BRAHMEGUPTA.

[*Circ.* B.C. 540—80 A.D.]

Sir J. Malcolm, in his "History of Malwa," describes the early history of this province as involved in darkness and fable; but he supposed Oojein to have had more undoubted claim to remote antiquity than any other city in India.

He says:—"We find in Indian manuscripts Malwa noticed as a separate province 850 years before the Christian Era, when Dhunjee, to whom a divine origin was given, restored the power of the Brahmins, which, it is stated, had been destroyed by the Buddhists, many remains of whose religion are still to be found in this part of India. In the excavation of a mountain near Baug we trace, both in the form of the temples and in that of the figures and symbols which they contain, the peculiar characteristics of the Buddhist worship." With regard to the date ascribed to Dhunjee, he remarks that "the principal Buddha is not so old as eight centuries before Christ," and that "his age has been accurately ascertained to be about five and a half centuries before Christ."

Now, it is remarkable, from their own statements, that a great error was committed by the Hindu writers of the period here referred to, for they have added together the genealogies of two distinct dynasties of different tribes, as if they were continuous in the same line (the princes of which were, for some time, contemporaneous). When this error is corrected, the date of Dhunjee is brought down from 850 B.C. to 109 B.C., and also places the era of Raja Bhoja (a great prince celebrated for the encouragement given by him to learning), at a mean date of 533 A.D.; whereas, the Hindu writers, by the error referred to, have placed him as living in the 10th and 11th centuries of our era.

In order to clearly present to the reader the nature of the error referred to, it is necessary only to allude to the four names, Dhunjee (stated by native writers to have existed at a period corresponding with 350 B.C.); Vicramaditya (56 B.C., on the same authority); Salivahana (at the era of the Saca, 79 A.D.), and Raja Bhoja (stated as living any time between 900 A.D. and 1100 A.D.). The writer proposes to show that, although the periods assigned to Vicramaditya and Salivahana are approximately correct, yet, by reason of the confusion between two dynasties, Dhunjee is placed at a period at least seven centuries too early, and Raja Bhoja is assigned to a date from five to seven centuries too late.

The corrected date agrees with nearly all the conclusions, regarding Raja Bhoja, which Colebrooke had arrived at on other grounds, although, to reconcile the inconsistencies in the Hindu accounts, he believed there had been many princes of the name of Raja Bhoja. (Essays, Vol. II., P. 53.)

This error of the Hindu writings will be apparent from the following account:—

In the "Summary of the History of the Princes of Malwa," which we find in the Ayeen Akberi, a series of tables is given of several dynasties of the Kings of Malwa.

These must have been furnished to Abul Fazel by learned men of his time, when he was compiling the "Institutes of Akber," and he appears himself to have visited Oojein. Now, the first of these tables, headed by Dhunjee (described as the chief of a tribe of the Deccan), comprises the names of five princes, who together are recorded to have reigned 387 years 7 months. The third prince on this list is Salivahana, and his two predecessors are said to have reigned altogether 186 years and 7 months. The era of Salivahana is, undoubtedly, the Saca, which in India is universally reckoned to be 78—9 A.D. The era of Dhunjee, according to this computation, would, therefore, be about 108 B.C., thus giving the date when Brahminism was re-established.

Table II (given in the "Ayeen Akberi") consists of 18 princes of the Punwar Dynasty, beginning with Adut Punwar, who, according to Sir J. Malcolm, was a Rajput, and the seventh on this list is Vicramaditya.

But we are told also by the Hindu writers that Salivahana made war upon Vicramaditya and took him prisoner, but granted his request that the Sambat, which is the era of Vicramaditya, and now universally admitted to be 56 B.C., should not be discontinued in public transactions. Nevertheless, Salivahana, on his accession to the throne, made use of another era ("Ayeen Akberi," Vol. I., p. 330).

In the same table are recorded the names of the remaining 12 princes beginning with Vicramaditya, inclusive, all being predecessors of Raja Bhoja, whose reigns altogether amount to 636 years and five months, which, reckoned from the beginning of the Sambat or 56 B.C., would place the reign of Raja Bhoja at about 580 A.D.

Abul-Fazel, however, states that Bowj (or Raja Bhoja) succeeded to the kingdom in the 541st year of the era "of Vicramaditya (*i.e.*, 485 A.D.), and that he made considerable additions to his dominions by conquest. His reign was celebrated for his justice and liberality, and he gave such encouragement to men of learning and wisdom that no less than 500 sages were to be found in his palace. He made trial of the abilities of them all, and found the most eminent amongst them were Beruj and Dhunpaul, whose compositions are highly esteemed to this day." (1595, "Ayeen Akberi." Vol. II., p. 55).

If this explanation of the error of Hindu writers be admitted, it would also explain how there should have arisen such great differences amongst European writers, regarding the important periods of the re-establishment of the religion of the Brahmins and the age of Bhoja, which, from the errors having been carried forward by

L

Hindu writers, has been placed anywhere between the 9th and 12th centuries, A.D.

The date which Abul-Fazel gives for the age of Raja Bhoja is certainly more to be relied upon than that deduced from the table of the Hindu manuscripts of Oojein, for in the court of the Emperor Akber, Abul-Fazel was surrounded by the most learned Hindus of his time, who were his assistants in compiling his "Institutes of Akber." In his researches regarding the Hindus, he could have no motive for altering the narratives and facts communicated to him, and he seldom resorted to conjecture.

He admits that he is not infallible, for he says:—"I had long set my heart upon writing something of the history of Hindostan, together with an account of the religious opinions of the Hindus. I know not if my anxiety herein proceeds from the love of my native country, or whether I am impelled by the desire of searching after truth, and relating matter of fact."

In his researches he is equally desirous with ourselves of obtaining trustworthy evidence regarding all that was related to him for the compilation of his work. He possessed advantages of obtaining information far superior to any that subsequent amateur antiquaries have ever enjoyed. He lived at least 200 years nearer the times when the incidents and facts of his information may have been better known, and, as the confidential minister of Akber, the high authority he held in the empire gave him the means of obtaining information on every side.

Few Indian names have excited more curiosity than that of Vicramaditya. In his court were assembled most of the learned men of his time; amongst them were the so-called nine gems, poets who were the ornaments of his court; one of them was the celebrated poet Calidas, author of a number of dramatic works, and other poems, in which are depicted the manners and customs of the age in which he lived; another was the distinguished lexicographer and poet, Amera-Sinha; his poems are said to have perished. In religion he was a Buddhist, and reputed to be a theist of tolerant principles. Hindu writers also give the names of several astronomers who were guests, but nothing further appears to have been related regarding them.

The two eras, Sambat (56 B.C.) and Saca (78—9 A.D.), are important in Hindu astronomy as marking the time when the era of Yudhisthira was discontinued or superseded by them.

Thus, the ancient astronomers, Parasara and Garga, employ the older era, and Colebrooke makes use of the argument that the astronomer Aryabhatta, about whose period there is some degree of uncertainty, since he does not make use of the Sambat of Vicramaditya nor the Saca era of Salivahana, but exclusively employs the epoch Mahabharata (that is, of Yudhisthira); therefore, he must have flourished before this epoch was superseded. Further, Davis seems to have held the opinion that before the older epoch was superseded there is evidence to show that the solar year began at the Winter Solstice, whereas the year of the Sambat of Vicramaditya begins at the Vernal Equinox.

For some centuries after the era of Vicramaditya, a period for the most part full of political and religious disturbances, from parties contending for supremacy, and from the expulsion of the Buddhists, which probably took place during this interval, little or nothing is known regarding the changes Hindu astronomy had undergone in its reconstruction.

At the epoch of the Saca, the Vernal Equinox, the moveable origin of the Solar Zodiac, was at least 7° to the east of the first of Aswini, the subsequently fixed "origin" of the Lunar Asterisms. Astronomers, who were at this time, doubtless, acquainted with and made use of rules of both the Lunar and Solar systems, must have found much inconvenience from this difference in position of the two points to which the longitudes were referred.

Other inconveniences must, also, have been felt by those astronomers who used the Solar Zodiac with a moveable origin, owing to

L 2

the different opinions held regarding the precession of the Equinoxes, the amount of which was stated differently by different authors.

It was a general opinion, as before stated, that the precession resulted from the libration of an Equinox, within limits of an assumed number of degrees, on each side of a mean fixed point. Of the various opinions cited by Colebrooke (Essays, p. 374, etc.), the one which gives the nearest to a correct value is that of Parasara. It is as follows:—"The same doctrine (of a libration) is taught in the 'Parasara Siddhanta,' as quoted by Muniswara; and if we may rely on the authority of a quotation of this author from the works of Aryabhatta, it was also maintained by that ancient astronomer; but, according to the first-mentioned treatise, the number of librations amounted to 581,709, and, according to the latter, of 578,159 in a calpa, instead of 600,000 (the number given in the Surya Siddhanta): and Aryabhatta has stated the limits of the libration as 24° instead of 27°."

Now, from the former statement, or that of Parasara, reckoning a complete libration (or, as it was by some authors called, revolution), of 4×24° or 96°, a mean annual precession of 46.53672" is deduced, and from that of Aryabhatta 46.2572" results, both of which are nearer the true value than that of the Surya Siddhanta of 54", which was adopted in all the other Siddhantas of modern astronomy.

About the year 480 A.D., the astronomer Varaha-Mihira is described as noting and seemingly expressing his surprise that the Solstices which, in the time of the Rishis, were, as recorded in former Sastras, the one in the middle of Aslesha, and the other in the first degree of Dhanishtha, should now, in his time, be, one in the first degree of Carcata, and the other in the first of Macra. This would imply that the Vernal Equinox was now, in his time, near the first of Aswini. He was of a sect which made use of the Solar Zodiac for expressing their longitudes, and he may have been one of those who perceived the advantage which would arise from making the beginning of the Solar Zodiac, Mesha, fixed and coincident with the Equinox when it was in the first of Aswini, and correcting the longitudes afterwards by merely adding the precession.

The astronomers of Oojein informed Dr. Hunter, who was in an embassy to that city, that there were two astronomers of the name of Varaha Mihira; to one of them they ascribed a date 122 Saca, corresponding to 200-1 A.D.; to another a date of 427 Saca, or 505-6 A.D. Sir W. Jones supposed, from astronomical data, Varaha Mihira to have lived about 499 A.D. Colebrooke, for similar reasons, supposed that, from two calculations, one placing him at 360 A.D., and the other at 580 A.D., he may have been living at the mean date of the two, or about 470 A.D.

Now, much of the authenticity and accuracy of our information regarding the more ancient astronomy of the Hindus depends upon the evidence that is derived from the numerous writings of Varaha.

He appears to have been of some astrological sect who had the Solar Zodiac for the foundation of their opinions, but he was familiar with the more orthodox doctrines of astronomers, who had the Nacshatras or Lunar Asterisms as the groundwork of their system.

It is, however, evident that he misunderstood and misrepresented some of their doctrines, such as in the contradictory opinions which he ascribed to Aryabhatta, and the strange doctrine before referred to, concerning the motion of the Rishis through each Asterism in one hundred years, an erroneous doctrine which was not held by the mathematical astronomers, Parasara, Brahmegupta, Bhascara, and other orthodox writers.

He is described as the author of a copious work on astrology, consisting of three parts, which he declared was abridged from earlier writers. It related to the computation of a planet's place (called Tantra); to lucky and unlucky indications (named hora);

and to prognostics regarding various matters, journeys, weddings, nativities, etc. (denominated Sacha). Of this work, the first section of the second part, known under the title of Vrihat-Jataca, comprising 26 chapters, is still extant.

The third part of this astrological work, containing 4,000 couplets, in 106 chapters, is also surviving, and is unimpaired, and known and cited, as the Vrihat Sanhita, or great course of astrology.

It is evident, however, that Varaha had recovered and had in his possession, a number of the ancient more orthodox astronomical works of the Hindus. It appears that he had the writings of Parasara, Garga, and Aryabhatta, which he commented upon. He was also the editor of five different orthodox works, entitled the "Pancha Siddhantica," a "knowledge of which he required, as requisites in the qualifications of an astronomer competent to calculate a calendar. Among other attainments, he required him to be conversant with time, measured by Yugas, etc., as taught in the five Siddhantas upon astronomy, named Paulisa, Romaka, Vasishtha, Saura, and Paita-Maha."

The "Pancha Siddhantica, as a complete work, edited by Varaha, has not been recovered; but the Saura, under the name of the Surya Siddhanta, is supposed to be entire, and the Paita-Maha is intended to mean the Brahma Siddhanta, and all the others are cited and assigned to different authors.

From all this it may be inferred that he admitted these ancient works to have been all established and referred to as great authorities in times long anterior to the date in which he was living; and from his references to Parasara and other Munis or Rishis, although he was mistaken with regard to some of their opinions, he had faith in the correctness of the observations recorded by them.

As before stated by the author of the Ayeen Akberi, at the court of Raja Bhoja in 485 A.D., there were assembled 500 sages. It may well be conceived that in this great assembly of learned men, there were present as honoured guests the most eminent mathematicians of the age. Varaha, and probably Jishnu, with his more celebrated son, Brahmegupta, also astronomers of another sect, followers of Aryabhatta, may have been present, maintaining the doctrines of their leader, in discussions with Brahmegupta.

In such assemblies mathematicians sought for distinction by propounding and giving solutions to difficult algebraical questions on astronomy. At the end of the 8th section of the 18th Chapter of the Brahma Siddhanta, translated by Colebrooke from the Sanscrit, about 60 questions on astronomy are proposed for solution. On concluding this chapter, Brahmegupta says :—

"These questions are stated merely for gratification; the proficient may devise a thousand others, or may resolve by the rules taught, problems proposed by others, as the sun obscures the stars, so does the proficient eclipse the glory of other astronomers, in an assembly of people, by the recital of algebraic problems, and still more by their solution.

"These questions recited under each rule with the rules, and their examples amount to a hundred and three couplets, and this chapter on the Pulverizer is the 18th."

Some few of these questions, with the methods which were adopted for their solution, will be given in a subsequent part of this work.

The name of Brahmegupta is held in the highest esteem by all Hindu writers. The age near which he lived is fairly well known. From certain observations which he made and recorded, Bentley calculated the date to be 535 A.D.; Colebrooke, from the position of the heavenly bodies observed by Brahmegupta, and allowing for uncertainty in inaccurate observations, was disposed to agree with Bentley, but assigned 581-2 A.D. as the result of his calculations. The astronomers of Oojein also gave 550 Saca, or 628 A.D as the date of Brahmegupta.

His most remarkable work was a revised and corrected edition of the ancient sacred work, the Brahma Siddhanta, from some earlier copy. This edition by Brahmegupta will be more particularly described hereafter.

This astronomer, as has been previously remarked, combatted the theory of his predecessor, Aryabhatta, concerning the rotation of the earth, and extracts from his arguments have been given.

Colebrooke remarks that Brahmegupta is more fortunate in refuting a theory of the Jainas, who, to account for the alternation of day and night, imagined that the daily changes were caused by the passage of "two suns and as many moons, and a double set of stars, and minor planets round a pyramidical mountain, at the foot of which is this habitable earth." His confutation of that absurdity is copied by Bhascara, who has added to it the refutation of another notion ascribed by Brahmegupta to the same sect respecting the translation of the earth in space, founded upon the idea that the earth, being heavy and without support, must perpetually descend.

The answer given to this is:—

"The earth stands firm by its own power without other support in space.

"If there be a material support to the earth, and another upholder of that, and again another of this, and so on, there is no limit. If, finally, self-support must be assumed, why not assume it in the first instance? Why not recognise it in this multiform earth?

"As heat is in the sun and fire, coldness in the moon, fluidity in water, hardness in iron, so mobility is in air, and immobility in the earth, by nature. How wonderful are the implanted faculties!

"The earth possessing an attractive force (like loadstone for iron, says the commentator on Bhascara), draws towards itself any heavy substance situated in the surrounding atmosphere, and that substance appears as if it fell. But whither can the earth fall, in ethereal space which is equal and alike on every side?

"Observing the revolution of the stars, the Buddhists acknowledge that the earth has no support; but, as nothing heavy is seen to remain in the atmosphere, they thence conclude that it falls in ethereal space.

"Whence dost thou deduce, O Baudda, this idle notion, that

because any heavy substance thrown into the air falls to the earth, therefore the earth itself descends? For, if the earth were falling, an arrow shot into the air would not return to it when the projectile force was expended, since both would descend. Nor can it be said that it moves slower, and is overtaken by the arrow, for heaviest bodies fall quickest (he supposes), and the earth is heaviest."

Abul Fazel gives a short but faithful account of the astronomy of the Hindus, as it was known by them during the past 18 centuries, and he relates a few instances regarding religious observances which seem to point to some sect of ancient priestly astronomers who held doctrines somewhat different from those of the Brahmins.

In the Ayeen Akberi (A.D. 1591, Vol. II., p. 69), it is narrated that:—

"In the Soobah (province) of Berar are many rivers, the principal of which is called the Gung-Kotemy (Gung, Gotuma), and sometimes the Godavery. The Hindus have dedicated this river to Kotum (*i.e.*, Gotuma or Buddha) in the same manner as the Ganges to Maha-Deva (Siva). They relate wonderful stories regarding it, and hold it in great veneration.

• • • • • •

"When the planet Jupiter enters the sign of Leo the people come from great distances to worship this river."

In page 164 it is related that:—

"In Kotehar (a place in Cashmere) there is a fountain which continues dry for eleven years, and when the planet Jupiter enters the sign of Leo, the water springs out on every Friday, but is dry all the rest of the week during the year."

This was probably a natural syphon, which was only allowed to flow every Friday during the year in which Jupiter was passing through the Constellation Leo.

Again, page 167:—

"Adjoining the Gurgong is a pass called Sowyuru at the extremity of which is a plot of ground measuring 10 Jerebs. When the planet Jupiter enters Leo, for a month's continuance, the soil of this place is so intensely hot that it destroys the trees, and if a kettle is set on the ground it will boil."

At page 353, referring to a place near the junction of the Jumna and the Ganges, Abul Fazel observes:—

"It is astonishing that when the planet Jupiter enters the Constellation Leo, a hill rises out of the middle of the Ganges and remains for a month, so that people go upon it and perform divine worship." Mr. Barlow was of opinion that this legend about a hill rising out of the Ganges would seem to indicate a low river owing to want of rain, and this might readily be associated with a failure of the harvest, and would hardly be forgotten if it occurred two or three times in succession.

At page 183:—

"In the reign of Raja Bunjir (Castunri), whilst the sun was in Leo, there was a fall of snow which totally destroyed the harvest and occasioned a terrible famine."

From the circumstance of so much importance having been attached to the entrance of Jupiter into the Constellation Leo, we may infer:—

First, that the astrological priesthood of the sect which made this specific circumstance a stimulus to the devotions of the people, had the Solar Zodiac for the foundation on which they based their astronomy.

Secondly, that they had at least one of the two Cycles of Jupiter, this being his year consisting of nearly 12 Saura years, the other being the Cycle of Vrihaspati, consisting of 60 years. Each of the 60 years was called Vrihaspatis Mana, or Madhyama, his mean motion through one sign. These years had each a separate name.*

* The cycle of Vrihaspati of 12 years, as described by Parasara, quoted by Varaha-Mihira, is thus explained.

"The name of the year is determined from the Nacshatra in which

Thirdly, that the Constellation Leo should have been deemed more sacred than any of the others, over each of which Jupiter occupied a year in the transit, is significant of the period, when the Solstice was in Leo; a propitious time, which was equally sacred to all the tribes of the East, as a time when sacrifices and prayers were ordained to be offered to the Supreme Being, the memory of

---

Vrihaspati rises and sets (heliacally), and they follow in the order of the lunar months."

"The years beginning with Cartic commence with the Nacshatra Critica, and to each year there appertain two Nacshatras, except the 5th, 11th and 12th years, to each of which appertain three Nacshatras."

There was a difference of opinion amongst the Astronomers regarding the naming of the years.

The names and order of the 12 Vrihaspati years were not the same as those of the cycle of 60.

According to Sasipura and others, the Nacshatra in which Jupiter rises gives the name to the year.

Casyapa says the name of the Samvatsura Yuga and the years of the cycle of 60 are determined by the Nacshatra, in which he rises, and Garga gives the same account. Some make the cycle to begin on the first day of the month of Chaitra, &c., whatever may be the Nacshatra in which Jupiter is. According to Parasaras' rule, which gives also the character distinguished as good or bad, with the names and order of the corresponding Nacshatras, they are set forth as follows, with the presiding deities.

Years.	Nacshatras.	Deities.	Characters.
Cartic	Critica, Rohini	Vishnu	Bad.
Agrahayan	Mrigasiras, Ardra	Surya	Bad.
Paush	Punarvasu, Pushya	Indra	Good.
Magh	Aslesha, Magha	Agni	Bad.
Phalgun	Purva Phalguni, U-Phalguni, Hasta	Twashta	Neutral.
Chaitr	Chitra, Swati	Ahivradna	Good.
Vaisach	Visacha, Anuradha	Pitris	Bad.
Jaishth	Jyeshthá, Mula	Viswa	Bad.
Ashar	P-Ashara, U-Ashara	Soma	Good.
Sravan	Sravana-Dhanishtha	Indragni	Good.
Bhadr	Satabhisha, P-Bhadrapada, U-Bhadrapada	Aswina	Good.
Aswin	Revati, Aswini, Bharani	Bhaga	Good.

The commentary states, "It is in the Soma Sanhita, that the presiding Devas are thus stated.

which circumstance had been preserved, although the Solstice had passed out of the Constellation Leo nearly 3,000 years before that time.

In the 3rd Volume of the "Asiatic Researches, Davis has given the results of an investigation regarding the cycle of 60 years of Vrihaspati, which some have supposed to be of Chaldean origin, or the Sosos. He says that it was wholly applied to astrology.

Davis was a civil servant of the East India Company at Bhaugulpore in 1789 and 1791, he gave two papers to the second and third volumes of the "Asiatic Researches," Calcutta.

The first on "The Astronomical Computations of the Hindus;" the second on "The Indian Cycle of 60 years."

In the latter paper, to illustrate some of his remarks, he publishes a plate of the Hindu ecliptic, of which plate annexed (page 157) is a reduced copy.

He says, "Its origin is considered as distant 180° in longitude from Spica; a star which seems to have been of great use in regulating their astronomy, and to which the Hindu tables of the best authority, although they differ in other particulars, agree in assigning six

---

In the cycle of 60 years are contained five cycles of twelve, which five cycles or Yugas, are named—

Samvatsara, over which presides			..	..	Agni.
Parivatsara	,,	,,	..	..	Arca.
Idavatsara	,,	,,	..	..	Chandra.
Anuvatsara	,,	,,	..	..	Brahma.
Udravatsara	,,	,,	..	..	Siva.

"The first year of the cycle of sixty, named Prabhava, begins when, in the month of Magha, Vrihaspati rises in the first degree of the Nachshatra Dhanishtha, because when Vrihaspati rises in 9^s, 23 20′ Surya (the Sun) must be 10^s 6° 12′."

Names are given to all the 60 years of Jupiter, beginning with Prabhava as the first. The order in which these years are arranged is given by Davis in the accompanying plate, which is copied from the one given by him in the third volume of the "Asiatic Researches."

signs of longitude, counting from the beginning of Aswini, their first Nacshatra."

In a preceding Chapter it has been stated that the beginning of the Nacshatra, the first of Aswini, was determined from a reference to it, in several Hindu works, by means of a star, which has been identified as ζ Piscium, and from the statement that when the vernal

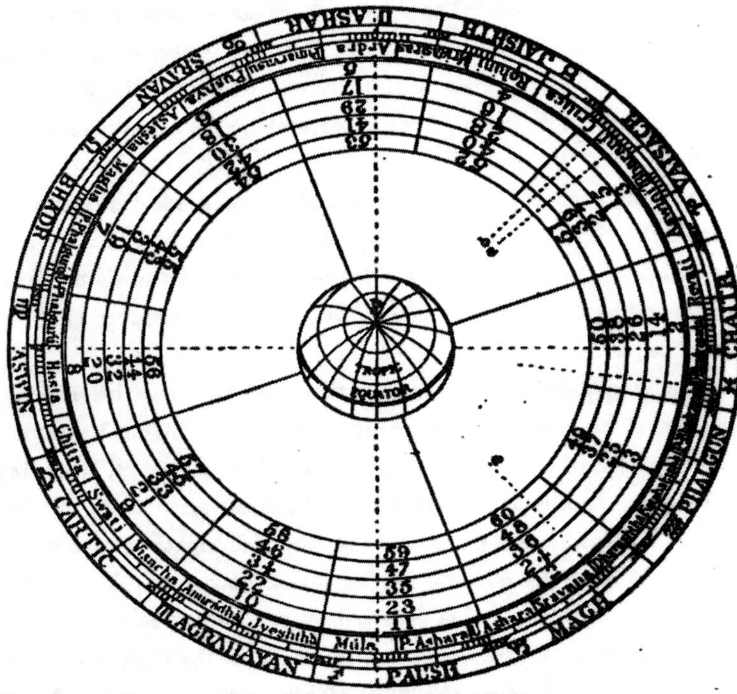

The Hindu Ecliptic and Cycles of Jupiter.—*According to Davis' "Asiatic Researches," Vol. III.*

equinox coincided with this star, the longitudes were afterwards reckoned as fixed, and that the time when this change was made was about 570 A.D.

The principal object of Mr. Davis' plate was to show the nature of the two cycles of Jupiter, the smaller cycle consisting of 12 Saura years, and the larger cycle comprising five of these, or 60 Saura years.

The line *a* passing through the beginning of Dhanishtha and the middle of Aslesha, shows the position of the solstitial colure about the year 1110 B.C., and the line *b* through the beginning of Critica represents the position of the equinoctial colure about the year 1305 B.C., whilst the other line *a*, passing through Bharani, shows the position of the latter colure when the solstitial was at the beginning of Dhanistha (1110 B.C.).

Davis observes that "The solar months correspond in name with the like number of Nacshatras; this is ascribed to the months having been originally lunar, their names derived from the Nacshatras in which the moon, departing from a particular point, was observed to be at the full."

The most ancient Indian year, named Saura, made a day equivalent to the time taken by the sun to pass through each degree of the Ecliptic; so that in one revolution of the sun the Saura year, corresponding to a motion of 360°, had the same number (360) of Saura days.

By analogy, Jupiter's period consisted of 12 times his Madhyhama, or "his mean motion through one sign." The Vrihaspati year consisted of one-twelfth of his period. To be exact, according to Laplace, Jupiter's Sidereal Period, in mean solar time, is 4,332 days 14 hours 18 mins. 41 sec., so that the Vrihaspati year would be 361 days 1 hour 11 min. 30 sec. of mean solar time. It was not, therefore, a period of either Saura days or of mean solar days, but of the time he took through each sign of 30° of the Ecliptic, and his cycle was complete when the 360° of his path was completed. It was called a cycle of 12-years, but they were neither solar years nor Saura years, for one or other of which they have sometimes been mistaken.

Davis was of opinion that as the year Cartic is always placed the first of the 12 Vrihaspati years, it may be inferred that there was a time when the Hindu solar year, as well as the Vrihaspati cycle of 12, began with the sun's arrival in or near the Nacshatra Critica.

He remarks that, "The commentator of the Surya Siddhanta

expressly says that the authors of the books generally termed Sanhitas, accounted the Deva day to begin in the beginning of the sun's Northern road; now the Deva day is the solar year, and the sun's Northern road begins in the Winter Solstice.

"This might, moreover, have been the custom in Parasara's time, for the phenomenon which is said to mark the beginning of the Vrihaspati Cycle of 60, refers to the beginning of Dhanishtha, which is precisely that point of the Ecliptic through which the Solstitial Colure passed when he wrote.

We are told by Hindu writers that the Buddhist sages were divided into six sects, that their works were upon various philosophical subjects; they had a treatise on logic, another on the folly of religious distinctions and ceremonies; they had a history of Buddhist philosophers, other works on the doctrines of Vrihaspati, and also a treatise on astrology.

Now, an astronomical or astrological work was known amongst the Hindus under the name of the Brihaspati or Vrihaspati Siddhanta. It is mentioned in the Ayeen Akberi among nine astronomical works of the Hindus, as one of four, which were supposed to have had a divine origin, and that this had been dictated by Jupiter, or Vrihaspati, whose name it bore. Hitherto this work has not been recovered.

That the name of the planet should have been associated with that of the sage Vrihaspati would seem to imply a connection of this astronomical work with the Buddhist religion, and that, in the Siddhanta of Vrihaspati would probably be found rules which regulated the observances of the Buddhist faith.

About this time it would appear that astrological sects had become partially recognised by the more orthodox astronomers, for some of their tenets, as also the Cosmogony of the Puranas, obtained a place alongside of the strictly mathematical doctrines of the Siddhantas, which they have since retained. But this concession does not extend to the absurd belief in the interposition of the dragon or monster Rahu in eclipses. Yet some of the devout Hindus cannot dispute the divine authority of the Purannas, and the learned try to evade the question, some of them saying that certain doctrines as stated in other sastras, "Might have been so formerly, and may be so still, but for astronomical purposes astronomical rules must be followed."

Nerasinha, in his commentary on the Surya Siddhanta, explains that "Rahu and Ketu, the head and tail of the monster, could only mean the position of the moon's nodes, and the amount of latitude on which eclipses depend," but he also says that "the belief in the actual presence of such a monster may be believed as an article of faith, without prejudice to astronomy."

CHAPTER XII.

ASTRONOMICAL WORKS OF THE HINDUS.

On the restoration of their power by the Brahmins, some time before the beginning of the Christian Era, their astronomical system was reconstructed, upon an orderly plan, and regular in its parts, the detached portions of more ancient treatises being collected, the rules being arranged and set forth in text books, without proof, and often without explanation.

The difficulties under which modern Hindu astronomers have laboured, and which have kept them in a stationary position with regard to their astronomy, as compared with the progress in other countries, may be well conceived when we consider the absence of printing. Consider the period in Europe, before the invention of printing, when accumulated knowledge existed principally in manuscripts, jealously guarded, loaned out, but protected by sufficient security for their due return. Had such a condition of things existed to the present time, what would have been the fate of astronomy, and of the mathematical sciences, in the absence of the Principia, the Mechanique Celeste, and the many other great mathematical works of the last century? Could we, indeed, have claimed any superiority of knowledge, under such circumstances, over that displayed by the Mediæval astronomers of India?

Of the eighteen or twenty ancient astronomical works referred to by ancient Hindu writers, under the name of Siddhantas, or "Established Conclusions," nine are mentioned by Abul Fasel in the Institutes of Akber, namely:—

1. Brahma Siddhanta.
2. Surya Siddhanta.
3. Soma Siddhanta.
4. Brihaspati Siddhanta.
5. Garga Siddhanta.
6. Narada Siddhanta.
7. Parasara Siddhanta.
8. Pulastya Siddhanta.
9. Vasishtha Siddhanta.

M

The names of others are those of—

10. Vyasa.
11. Atri.
12. Kasyapa.
13. Marichi.
14. Manu.
15. Angiras.
16. Lomasa.
17. Pulisa.
18. Yavana.
19. Bhrigu.
20. Chyavana.

The first four are reputed by the Hindus to be inspired; the first supposed to have been revealed by Brahma, the second by the sun, the third by the moon, and the fourth by Jupiter. All the others are supposed to have been dictated by mortals, and of these few are now extant, being principally known by citations from mathematical writers.

An astronomical work not mentioned in the above list, but deservedly held in great esteem, is the Siddhanta Siromani of Bhascara Acharya, of comparatively more recent date (1150 A.D.)

Abul Fazel states that in his time (about 1550 A.D., he being murdered by banditti in 1603) "there were no fewer than eighteen different opinions regarding the various changes and creations which the Universe has undergone, and he mentions the traditions with reference to three of these opinions, the third of which he says was the most generally received, and that it was one related in the Surya Siddhanta, a work supposed to have been compiled at the close of the Sutya Yuga (Krita Yuga).

"To this day (he remarks) all the astronomers of Hindostan rely entirely upon this book."

It is impossible to say which of the two principal astronomical works, the Brahma Siddhanta or the Surya Siddhanta, is the more ancient, or when either of the original works were composed or compiled, for both have undergone revision at different periods of Indian history.

The difference between the two works is shown by a difference in the modes of explanation, and in the subjects treated by them.

As before mentioned, a revised version of the Brahma Siddhanta was edited by Brahmegupta, under the title of the Brahma Sphuta (or amended) Siddhanta, at some date between 530 and 580 A.D.

Nrisinha, the commentator of the Surya Siddhanta, affirms that Brahmegupta's rules are framed from the Vishnu Dharmottara Purana, in which the Brahma Siddhanta is contained. Various other works of the same name are referred to as being anterior to the revised edition of Brahmegupta; such, for instance, as the Brahma Siddhanta of Sacalya, which is understood to have been one of the five systems from which Varaha Mihira compiled his Pancha Siddhantica.

According to Colebrooke, the Brahma Sphuta Siddhanta, as an entire work, consisted of 21 chapters. Of these, the arrangement of the subjects, from the 1st to the 10th, would appear to have been nearly the same as those of the corresponding chapters of the Surya Siddhanta. The 1st and 2nd consisted of the computation of the mean motions, and true places, of the planets. The 3rd contained the solution of problems concerning time, the points of the horizon, and position of places. The 4th and 5th set forth the calculation of lunar and solar eclipses. The 6th the rising and setting of the planets. The 7th, the position of the moon's cusps. The 8th, observations of altitudes by the Gnomon. The 9th, conjunctions of the planets. The 10th, their conjunction with the stars. The next ten are supplementary, including five chapters of problems, with their solutions; and the 21st explains the principle of the astronomical system in a compendious treatise on spherics, treating of the astronomical sphere and its circles, the construction of sines, the rectification of the apparent place of the planet from mean motions, the cause of lunar and solar eclipses, and the construction of the armillary sphere.

The copy in the possession of Colebrooke was defective, wanting the 6th, 7th and 8th chapters, with gaps of greater or less extent in

M 2

the preceding five; it was supposed to have been transcribed from an exemplar equally defective.

The 11th is represented as a curious chapter containing a revision and censure of earlier writers.

The five chapters from the 13th to the 17th, inclusive, relate to problems concerning the mean and true places of the planets, times, points of the horizon, and other matters impossible to specify on account of defects in the 16th and 17th chapters. The 19th, 20th and 21st were wanting. One of them, from references to it, appears to have treated of time, solar, sidereal, lunar, etc.; another, from like references, treated of the delineation of celestial phenomena by diagrams.

It is a matter of regret that Colebrooke has only given translations of the 12th and 18th chapters of this work.

The 12th chapter of the Brahma Siddhanta consists of rules for the solution of questions in arithmetic, algebra, and mensuration, together with a few examples applied to problems.

To show the nature of the subjects which are treated in this chapter, the following is a description of the contents, taken from the words of the rules, and put in a modern form.

The questions on geometry are mostly intended to be answered arithmetically; but the rules are usually applicable to general solutions, suggesting an algebraical origin.

BRAHMA-SPHUTA SIDDHANTA OF BRAHMEGUPTA.

Summary of contents of the 12th Chapter—Ganitadhya.

The rules are expressed in words, but to show their import they are here set forth in the modern algebraical form.

Reduction of fractions to a common denominator.

Addition and subtraction of fractions.

Reduction of mixed numbers to common vulgar fractions.

Multiplication and division of fractions.

Involution and Evolution.

$(a+b)^3 = a^3 + 3a^2b + 3ab^2 + b^3$ expressed in words, preparatory to the extraction of the cube root of a number.

The rule of three—Interest, Partnership, Barter.

Arithmetical Progression.

$$S = (a+l)\frac{n}{2};\ l = a+(n-1)d;\ n = \frac{d-2a+\sqrt{(2a-d)^2+8sd}}{2d}$$

SERIES.

$$1+3+6+10+\ \ldots\ \frac{n(n+1)}{2} = n(n+1)(n+2).]$$

$$1^2+2^2+3^2+\ \ldots\ n^2 = \frac{n(n+1)(2n+1)}{6}.$$

$$1^3+2^3+3^3+\ \ldots\ n^3 = \frac{n^2(n+1)^2}{4}.$$

MENSURATION : PLAIN FIGURES.

Area of a triangle $= \sqrt{s(s-a)(s-b)(s-c)}$.

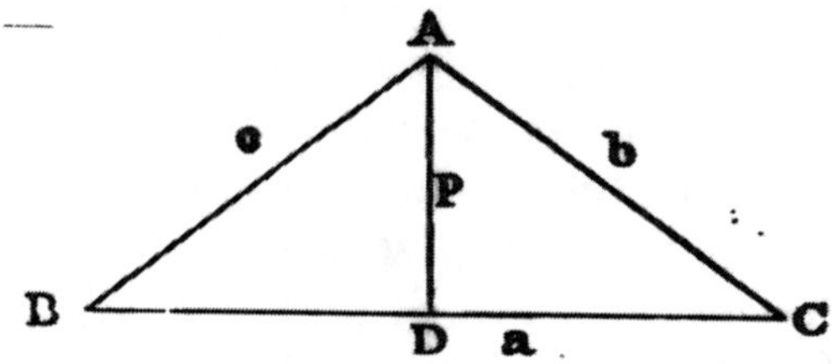

Perpendicular on a base, $p = AD = \sqrt{\left\{c^2 - \frac{1}{4}\left(a+\frac{b^2-c^2}{a}\right)^2\right\}}$

Segments of base, $BD = \frac{1}{2}\left(a+\frac{b^2-c^2}{a}\right)$

$$CD = \frac{1}{2}\left(a-\frac{b^2-c}{a}\right)$$

Radius of circumscribing circle $= \dfrac{bc}{2p}$.

52. The half-day being divided by the shadow (measured in lengths of the Gnomon) added to one, the quotient is the elapsed, or the remaining portion of the day, morning or evening.

The half-day, divided by the elapsed, or remaining portion of the day, being lessened by subtraction of one, the residue is the number of Gnomons contained in the shadow.*

53. The distance between the foot of the light and the bottom of the Gnomon, multiplied by the Gnomon of given length, and divided by the difference between the height of the light and the Gnomon, is the shadow.

A light at A is supposed to cast a shadow D B of the gnomon G.

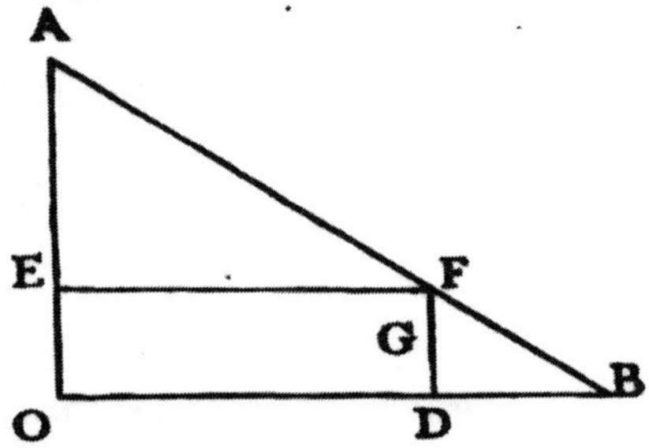

By proportion in the two triangles D B F and E F A $\frac{DB}{G} = \frac{EF}{AE}$ $DB = \frac{OD \cdot G}{AO - G}$.

54. The shadow multiplied by the distance between the tips of the shadows, and divided by the difference of the shadows, is the base. The base, multiplied by the Gnomon and divided by the shadow, is the height of the flame of light.

Here—O C = Base, C B = difference between tips of shadows.

$DB = S_1$, $EC = S_2$, $AO = H$.

$$OC = \frac{S_2}{S_1 - S_2} \times CB, \quad OA = \frac{G}{S_2} \times OC.$$

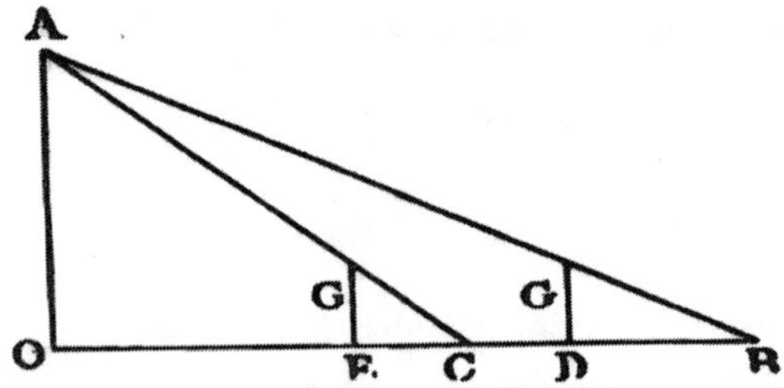

* This rule is considered by Crishna (a commentator on this Siddhanta) to have been copied from earlier writers, and useless.

By proportion—

$$\frac{AO}{G}=\frac{OB}{DB}=\frac{OC}{EC}\therefore\frac{OB}{OC}=\frac{DB}{EC}\text{ and }\frac{OB-OC}{OC}=\frac{DB-EC}{EC}$$

$$\therefore\frac{CB}{OC}=\frac{DB-EC}{EC},\text{ hence }OC=\frac{EC.CB}{DB-EC}=\frac{S_2.CB}{S_1-S_2}$$

$$AO=OC.\frac{G}{EC}=\frac{G.OC}{S_2}.$$

Heights and Distances.

Ex. On the top of a hill live two ascetics. One, being a wizard, travels through the air, springing from the summit of the mountain, he ascends to a certain elevation and proceeds by an oblique descent diagonally to a neighbouring town. The other walks down the hill, goes by land to the same town. Their journeys are equal. I desire to know their distance from the hill, and how high the wizard rose.

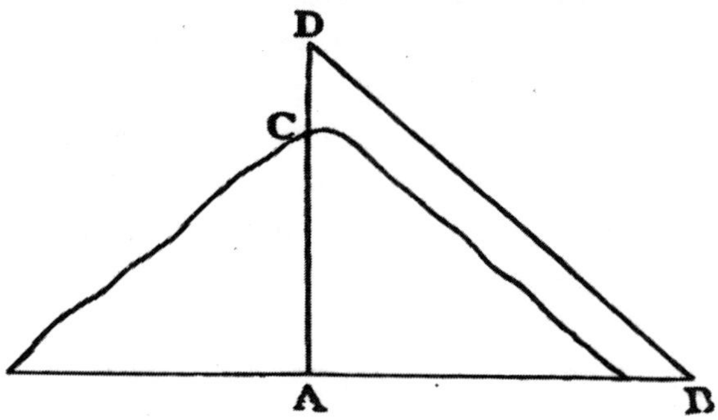

Let –

$AB=a$, $AC=b$, $CD=h$.

By supposition, $a+b=h+\{(b+h)^2+a^2\}^{\frac{1}{2}}$.

Or, $a+b-h=\{(b+h)^2+a^2\}^{\frac{1}{2}}$.

$\therefore\ 2ab=2h(a+b)+2bh$.

Or, $h=\dfrac{ab}{a+2b}$, let $a=mb$, m being any arbitrary number.

Then $h=\dfrac{m}{m+2}.b$.

The 18th chapter of the Brahma Sphuta (amended) Siddhanta is a treatise principally on the solution of indeterminate equations, employed partly in questions relating to abstract numbers, but mostly in the solution of problems in astronomy, relating to the sun, the moon, and the planets, with their nodes, apogees, etc.

It begins by stating that questions can scarcely be solved without the "pulverizer," and a rule is propounded for its investigation. This, however, from some cause, is a rule for a different purpose, connected with calculating a cycle or Yuga of three or more planets by means of the "pulverizer"; but is left unexplained.

The rule for the formation of the "pulverizer" is thus given:—

Rule 3-6.—The divisor which yields the greatest remainder is divided by that which yields the least; the residue is reciprocally divided, and the quotients are severally set down one under the other. The residue [of the reciprocal division] is multiplied by an assumed number such, that the product having added to it the difference of the remainders, may be exactly divisible [by the residue's divisor]. That multiplier is to be set down [underneath], and above it, and the product, added to the ultimate term, is the Agranta. This is divided by the divisor yielding the least remainder, and the residue, multiplied by the divisor yielding the greatest remainder, and added to the greater remainder, is a remainder of [division by] the product of the divisors.

Thus may be found the lapsed part of a Yuga of three or more planets by the method of the "pulverizer."

It may, however, be gathered, from examples given to different rules, that the method of the "pulverizer," in some respects, resembles one of the two rules usually given in modern works of algebra, for the solution of the indeterminate equation.

$$\frac{ax+c}{b} = y \text{ in integers,}$$

in which a is called the multiplier, b the divisor, and c the additive.

The fraction $\frac{a}{b}$, here supposed to be in its lowest terms, after division by the greatest common measure, is subjected to the reciprocal division of the numerator and denominator, and the several quotients are placed in order one above another. So far, the operation is the same in both methods; the method of our algebra then supposes the formation of a series of fractions converging to $\frac{a}{b}$, of which the last convergent may be assumed to be $\frac{p}{q}$. The Indian method proceeds to seek this convergent by operating backwards, upwards through the several partial quotients, the constant c being involved in each step, so that finally the result becomes . . $\frac{cp}{cq}, \frac{a}{b}$, and we have $a \,.\, cq - b \,.\, cp = c$, but by supposition $a \,.\, x \pm b\,y = c$, and the subsequent steps may be supposed to be identical with those of the modern methods of solution. $x = bm + cq$, $y = am + cp$, where m is any arbitrary number.

As applied to astronomical questions, a is the number of revolutions of a planet in a calpa; b the number of days in a calpa; and, c is called the residue of revolutions. $\frac{c}{b}$ the fraction of a revolution may be expressed in signs, degrees, and minutes.

For the purpose of explaining a few extracts, examples of computation from the Brahma Siddhanta are here stated. A rule for the calculation of the place of a planet is given, with the table to which it refers, and also a brief table of the same import with smaller numbers by Brahmegupta, which, instead of the inconveniently large numbers of the other table, were employed for illustration and for giving instruction to students of astronomy in the methods of calculation.

To find the mean places of the planets at a given midnight at Lanka:—

Rule.—Multiply the number of elapsed days by the number of revolutions of the planet in a calpa, and divide the product by the number of days in a calpa. The quotient will be the elapsed revolutions, signs, degrees, etc., of the planet (or longitude from 1st Aswini).

Mean motions of the Sun, Moon and Planets in a Calpa.

	Revolutions in a Calpa according to Brahmegupta.
Sun	4,320,000,000
Moon	57,753,300,000
Mars	2,296,828,522
Mercury ...	17,936,998,984
Jupiter ...	364,226,455
Venus	7,022,389,492
Saturn... ...	146,567,298
Days in a Calpa	1,577,916,450,000

For facility's sake, the revolutions and days are put as follows:—

Revolutions in least terms.		Days in least terms.
3	... ...	In 1,096 days.
5	... ...	„ 137 „
1	... ...	„ 685 „
13	... ...	„ 1,096 „
3	... ...	„ 10,960 „
5	... ...	„ 1,096 „
1	... ...	„ 10,960 „
☽'s Apogee	...	„ 2,740 „
☽'s Node	...	„ 5,480 „

"(7.) Question 1.—He who finds the cycle (Yuga) and so forth, for two, three, four or more planets, from the elapsed cycles of the several planets given, knows the method of the 'pulveriser.'

"Example.—What number divided by six has a remnant of five; and divided by five, has a residue of four; and by four, a remainder of three; and three, of two?"

The answer given to this question by Brahmegupta is 59.

But the question given, being indeterminate in form, there are many other answers. The general solution is:—

$N = 60n - 1$, where n is any arbitrary integer.

Thus when $n = 1$, N	59
$= 2$	$= 119$
$= 3$	$= 179$
$= 4$	$= 239$, etc., etc.

All of which satisfy the conditions. No doubt Brahmegupta was aware of this.

"(9.) Question 2.—He who deduces the number of [elapsed] days from the residue of revolutions, signs, degrees, minutes, or seconds, declared at choice, is acquainted with the method of the 'pulveriser.'

"Example.—When the remainder of solar revolutions is eight thousand and eighty, tell the elapsed portion of the calpa, if thou have skill in the 'pulveriser.'"

The answer given is 1,000 days.

To show how this arises, taking the solar revolutions from the smaller table, as 30, and the corresponding days as 10,960, and assuming x as the required number of elapsed days,

$$\frac{30x+8080}{10960} = \frac{3x+808}{1096} = y. \quad \text{suppose,}$$

y being the number of complete revolutions in x days, from which the general solution is

$$x = 1096\,m - 96, \text{ and } y = 3\,m - 1, \; m \text{ being arbitrary.}$$

If $m = 1$	$x = 1000$ days,	$y = 2$ revolutions.	
$= 2$	$= 1196$ „	$= 5$ „	
$= 3$	$= 1292$ „	$= 8$ „	etc., etc.

"(10.) Example.—To what number of elapsed days does that amount of hours correspond, for which the residue of lunar revolutions arising is four thousand one hundred and five?"

The answer deduced is 821 hours.

Taking the revolutions of the moon from the smaller table, as 5, and the corresponding days 137, we have, reducing the days into hours, of which there are 60 to the day,

$$\frac{5x-4105}{60 \times 137} = y,$$

where, as before, x represents the elapsed time, and y the number of complete revolutions; from which we have

$$x = 821 + 1644\,.\,y,$$

$$\therefore \text{When } y = 0,\; x = 821,$$

$$\text{„} \quad y = 1,\; x = 2465, \text{ etc., etc.}$$

THE SURYA SIDDHANTA.

Of all the astronomical works, however, of the Hindus, the one which claims particular attention is the Surya Siddhanta, as before stated.

Nothing authentic is known of the compilers of this work, much less of its composers, although considerable speculation has existed regarding its origin. Colebrooke says:—"Both Varaha-Mihira and Brahmagupta speak of a Saura (or Surya) Siddhanta, which is a title of the same import. Again, more than one edition of a treatise of astronomy has borne the name of Surya (with its synonym, the sun); for Lacshmidasa cites one under the title of Vrihat Surya-Siddhanta—in that commentator's opinion, and consistently with his knowledge, more than one treatise bearing the same name existed."

Colebrooke, when discussing the question of the antiquity of the Surya Siddhanta with Bentley, admitted generally the position:—"That the date of a set of astronomical tables, or of a system for the computation of the places of planets, is deducible from the ascertainment of a time when that system, or set of tables, gave results nearest the truth, and granting that the date mentioned (about 928 A.D.) approximates within certain limits to such an ascertainment—the book which we have now under the name of Surya or Saura Siddhanta may have been, and probably was, modernised from a more ancient treatise of the same name."

No certain information is derived from internal evidence regarding its origin. The work itself states that at the end of the Krita Yuga, a great demon named Maya is called upon by a man of divine origin, to listen to a discourse upon the science of astronomy, which had undergone some changes from what it had been in more ancient times, in these words:—

"Hear attentively the excellent knowledge which the sun himself formerly taught to the great saints in each of the Yugas.

"I teach you the same ancient science which the sun himself formerly taught. The difference (between the present and the ancient works) is caused only by time, on account of the revolution of the Yugas."

The commentator adds:—"Arca (the sun), addressing Maya, who

attended with reverence, said, 'Let your attention, abstracted from human concern, be wholly applied to what I shall relate. Surya, in every former Yuga, revealed to the Munis, the invariable science of astronomy. The planetary motions may alter, but the principles of the science are always the same.'"

It may be supposed that Meya represents some early disciple, to whom the Acharya or teacher is giving instruction, and after thus calling his attention to the subject, he proceeds to the distinction between two kinds of time, one of which is measurable, the other immeasurable. The measurable is that to which mortals are limited, the immeasurable is that which may approximate to the infinitesimally small, or to the infinitely great. In calculation it not infrequently happens that small fractions of time arise, which are required to be expressed with perfect accuracy, and as this is not always possible, different methods are adopted for approximating to a true value. Since the introduction of decimal fractions, the more ancient divisions of time in Europe are being gradually abandoned. In Eastern countries tables of time include not only days, hours, minutes and seconds, but also thirds, fourths, etc., and the sexagesimal division of the day into 60 Ghaticas (Indian hours), the Ghatica into 60 Palas, and a Pala into six Pranas, has some analogy to the sexagesimal division of the circle, in the six times sixty degrees of the Ecliptic, the degree into sixty minutes, the minute into sixty seconds, and so on to thirds, fourths, etc.

The ancient cycle of 60 years common to the Chaldeans, the Chinese, and the Hindus, consisted of five of Jupiter's periods of revolution, each of which consists of twelve solar years nearly.

From analogy, it might appear that the cycle of five solar years of the Vedic Calendar, as described in the 1st volume of Colebrooke's Essays, may have been adopted from its consisting of nearly five times twelve, or 60 synodic periods of the moon.

In the first chapter of the Institutes of Menu, which, according to Sir W. Jones, is one of the oldest Sanscrit books after the Vedas, there is a table or arrangement of infinite time, expressed in divine years, of which the following is an extract:—

"The sun causes the division of day and night, which are of two sorts, those of men, and those of the Gods; the day for the labour of all creatures in their several employments; the night for their slumber.

"A month is a day and night of the Patriarchs, and it is divided into two parts; the bright half is their day for laborious exertions; the dark half their night for sleep.

"A year is a day and night for the Gods, and that, also, is divided into two parts; the day is when the sun moves towards the North; the night when it moves towards the South. Learn now the duration of the day and night of Brahma, with that of the ages, respectively and in order.

"Four thousand years of the Gods they call Krita (or Satya) Age; and its limits at the beginning and at the end are in like manner as many hundreds.

"In the three successive ages, together with their limits, at the beginning and end of them, are thousands and hundreds diminished by one.

"This aggregate of four ages, amounting to twelve thousand divine years, is called an age of the Gods; and a thousand such divine ages added together must be considered as a day of Brahma. His night has also the same duration.

"The before-mentioned days of the Gods, or twelve thousand of their years, multiplied by seventy-one, form what is named here a Manuwantara. There are alternate creations and destructions of worlds through innumerable Manuwantaras. The Being supremely desirable performs all this again and again."

A similar arrangement is given in astronomical works of later times, but with the ages reckoned in years of mortals, and, for comparison, the following extract is taken from the Surya Siddhanta:—

Extract from the Surya Siddhanta.

"A solar year consists of twelve solar months; and this is called a day of the Gods.*

"An Ahoratra (day and night) of the Gods and that of the Demons are mutually at the reverse of each other (viz., a day of the Gods is the night of the Demons, and, conversely, a night of the Gods is a day of the Demons). Sixty Ahoratras multiplied by six make a year of the Gods and of the Demons.

"The time containing twelve thousand years of the Gods is called a Chatur Yuga (the aggregate of the four Yugas, Krita, Treta, Dwapara, and Kali).

"These four Yugas, including their Sandhya and Sandhyansa,† contain 4,320,000 years.

"The tenth part of 4,320,000, the number of years in a Great Yuga, multiplied by 4, 3, 2, 1, respectively, make up the years of each of the four Yugas, Krita and others, the years including their own sixth part, which is collectively the number of years of Sandhya and Sandhyansa (the periods at the commencement and expiration of each Yuga).

"According to the technicality of the time called Murta, 71 Great Yugas (containing 306,720,000 solar years) constitute a Manuwantara (a period from the beginning of a Manu to its end), and at the end of it, 1,728,000, the whole number of (solar) years of the Krita, is called its Sandhi; and it is the time when a universal deluge happens.

"Fourteen such Manus, with their Sandhis, constitute a Kalpa,

---

* The Gods are supposed to reside on Mount Meru under the North Pole, where the day lasts for six months.

The Demons are said to reside at the South Pole.

† Sandhya and Sandhyansa are the Dawn and Evening Twilight, and as the days of mortals have these, so also from analogy those of the Gods had them likewise.

at the beginning of which is the fifteenth Sandhi, which contains as many years as a Krita does.

"Thus, a thousand of the Great Yugas make a Kalpa, a period which destroys the whole world. It is a day of the God Brahma, and his night is equal to his day.

"And the age of Brahma consists of a hundred years, according to the enumeration of day and night. One-half of his age has elapsed, and this present Kalpa is the first in the remaining half of his age."

In these tables of long periods of time, the age of the Gods of 12,000 divine years, when multiplied by 360, the number of Saura days in the ancient Saura years, becomes 4,320,000 years, or the Maha Yuga. For a divine year is 360 years of mortals; and thus a day of Brahma of 1,000 divine ages, becomes 4,320,000,000 years of mortals, named the Kalpa.

Both these numbers have been received, in modern times, with much curiosity, and sometimes with abuse, when mistaken for numbers supposed to exaggerate periods of chronology.

They are, however numbers, which were adapted for the purpose of facilitating astronomical calculations, and they admit of a rational explanation.

THE MAHA YUGA.

First—With reference to the Maha Yuga:—

The most ancient sidereal year, both in India and in Chaldea, was assumed to consist of 360 Saura days.

The ancient Saura day is the variable time which the sun takes in its motion over each degree of the Ecliptic, the aggregate being the same as the number of parts into which the circle of the Ecliptic is divided; and from this, the apparent sidereal revolution of the sun, or the sidereal year is 360 Saura days.

But the Hindu astronomers also reckoned the sidereal year in mean solar time to be 365 days 6 hours 12 min. 36 sec., according to Pulisa, or, as a mixed number, $= 365\frac{207}{800}$ mean solar days.

Now, the absolute time of the apparent revolution of the sun in its orbit, or the sidereal year, is the same for both.

∴ A divine day = 360 Saura days = $365\frac{31}{120}$ mean solar days.

A divine year = 360 Saura years = 360 × $365\frac{31}{120}$ mean solar days.

12,000 deva years = 12,000 × 360 Saura years, or
The Maha Yuga = 4,320,000 Saura years
} = 1,577,917,800 mean solar days.

In the Surya Siddhanta the days of the Maha Yuga are reckoned to be 28 days more than in the Pulisa Siddhanta, *i.e.*, the Maha Yuga = 4,320,000 years = 1,577,917,828 days.

These two large integers, or other integers which are taken as equivalent to them, are fundamental in Indian calculations, which relate to the positions of the planets, such as their longitudes, times, from the epoch conjunctions, and oppositions, etc., etc.

The Maha Yuga, as a constant, is the same number in all Siddhantas, but the number representing the days in a Maha Yuga is slightly different in some of them. The above number of days is that which is given in the Pulisa and some other Siddhantas. In the Surya Siddhanta there are 28 days more, which would make a difference of only the fraction of a second in the length of the year, when divided among so many millions. In the Brahma Siddhanta, however, the number of days is given fewer by 1350 than the above, which would make the year less by about 27 seconds, and in the Arya Siddhanta it is made less by about six seconds.

As compared with European estimates of the sidereal year, those of the Siddhantas are all nearly four minutes too great.

The subjoined table from Colebrooke's Essays (Vol. II., p. 416) shows the number of revolutions made by the planets in a Maha Yuga, as specified in several Siddhantas: –

	According to Pulisa, quoted by Bhattotpala, in a Maha Yuga.	According to the Surya Siddhanta in a Maha Yuga.	According to Brahmegupta in a Calpa.
Sun	4,320,000	4,320,000	4,320,000,000
Moon (Periodical) ..	57,753,336	57,753,336	57,753,300,000
Mars	2,296,824	2,296,832	2,296,828,522
Mercury	17,937,000	17,937,060	17,936,998,984
Jupiter	364,220	364,220	364,226,455
Venus	7,022,388	7,022,376	7,022,389,492
Saturn	146,564	146,568	146,567,298
Days in a Maha Yuga or Calpa	1,577,917,800	1,577,917,828	1,577,916,450,000

It will be seen that they are not entirely in accord with each other. This want of harmony is accounted for by changes which these elements undergo in long periods of time, making it necessary from time to time to introduce corrections (or Bija). The commentator of the Surya Siddhanta says:—"The variation in the planetary motions is mentioned in the Vishnu Dhermotter, which directs that the planets be observed by an instrument whereby their agreement or disagreement may be determined in regard to their computed places; and in case of the latter, an allowance of Bija accordingly made."

The commentator Narasinha remarks that Vasishtha in his Siddhanta also recommends this occasional correction of Bija, and that it was the practice with Aryabhatta, Brahmegupta and others . . . . He says:—"A Ganita Sastra, whose rules are demonstrable, is true; and when conjunctions, oppositions, and other planetary phenomena, calculated by such Sastras, are not found to agree with observation; a proportionable Bija may be introduced without any derogation from their credit." . . . . .

And again:—"If a planet's place computed both by the Surya Siddhanta and the Parasara Siddhanta should be found to differ, which rule must be received as right? I answer, that which agrees with his place by observation, and the Munis gave the same direction."

The celebrated mathematician and astronomer Ganesa mentions that the Grahas (planets) were "right in their computed places, in the time of Brahma, Acharya, Vasishtha, Casyapa, and others, by the rules they gave, but in length of time they differed. . . . In the beginning of the Kali Yuga, Parasara's book answered, but Aryabhatta, many years afterwards, having examined the heavens, found some deviation, and introduced a correction of Bija. After him, when further deviations were observed, Durga Sinha, Mihira, and others made corrections. After them came the son of Jishnu Brahmegupta, and made corrections. Afterwards Cesava settled the places of the planets; and, sixty years after Cesava, his son, Ganesa, made corrections."

A similar table (of planetary revolutions) is given in our modern works of astronomy, the difference being that the periodic time is for one sidereal revolution of a planet instead of the time (a Maha Yuga) for a great number of revolutions of the same planet. Thus the sidereal period of one revolution of the earth is given in modern works as 365.2563744 mean solar days, a mixed number consisting of an integer 365 and a fraction carried to seven places of decimals. But if this mixed number be reduced to a vulgar fraction it becomes $\frac{3652563744}{10000000}$, which means that in ten millions of revolutions of the earth, or ten millions of sidereal years, there are 3,652,563,744 mean solar days. Thus we, by the use of a decimal point, express precisely what the Indian mathematicians meant to convey by the use of their system of large integral numbers.

The use of the great numbers (4,320,000 years, or 1,577,917,828 days), representing the years and days in a Maha Yuga, and the corresponding number of revolutions described by each of the planets in that time, might be exemplified in a variety of cases; but one or two examples will be sufficient here. They will illustrate the ease with which such calculations are made. Other examples as proposed in some of the Siddhantas have been already given.

Using the subjoined table, formed from the words by which they are expressed in the Surya Siddhanta:—

	Number of revolutions in a Great Yuga.	
The Sun	4,320,000	
Mercury	17,937,060	
Venus	7,022,376	
Mars	2,296,832	
Jupiter	364,220	
Saturn	146,568	
The Moon	57,753,336	
and ∴	53,433,336	Synodic revolutions.
The Moon's Apogee	488,203	
" Node	232,238	

	Number of days in a Great Yuga.
Sidereal days	1,582,237,828
Solar days	1,577,917,828
Lunar days	1,603,000,080

Let it be required to determine the number of revolutions, and parts of a revolution, made by the moon in a year.

In the column of the table Surya Siddhanta, the number of revolutions of the moon in a Maha Yuga is given, 57,753,336; dividing this number by 4,320,000, the years in a Maha Yuga, and in the successive divisors, omitting the factors 12, 30, 60, we have

4,320,000)57,753,336(13 revolutions,
56,160,000

360,000)1,593,386(4 signs,
1,440,000

12,000)153,336(12°,
144,000

200)9,336(46′,
9,200

$\frac{136}{200} = \frac{17}{25}{}'$.

That is to say, this makes 13 revolutions 4 signs 12° 46[illegible]′ in one year.

As a second example, let it be required to find the length of the sidereal year, from the days in a Maha Yuga. Reversing the process, and dividing the days by the apparent revolutions of the sun, and omitting in succeeding divisors the factors 24, 60 and 60 we have

```
4,320,000)1,577,917,800(365 days,
          1,576,800,000
          -------------
    180,000)1,117,800(6 hours,
            1,080,000
            ---------
          3,000)37,800(12 minutes,
                36,000
                ------
              50)1,800(36 seconds.
                 1,800
```

The sidereal year = 365 days 6 hours 12 minutes and 36 seconds.

ON THE KALPA.

The peculiar form in which the construction of the Kalpa is expressed attracted much attention more than a hundred years ago, and various theories were put forward to account for it.

Le Gentil had discovered from astronomical tables of Tirvalore that the Hindus made the value of the precession of the Equinoxes 54″, and this value is also assumed in all the modern Siddhantas. Sir W. Jones suspected that a more correct value of the precession had been obtained at some earlier period than that in which the Surya Siddhanta was compiled, and that it had a connection with the 14 Manuwantaras. He says:—"We may have reason to think that the old Indian astronomers had made a more accurate calculation, but concealed their knowledge under the veil of 14 Manuwantaras, 71 divine ages, etc."

After referring to the relapse of the astronomers into error without apparent cause, he concludes his remarks thus:—"Now, as it is hardly possible that such coincidences should be accidental, we may hold it nearly demonstrated that the period of a divine age (4,320,000 years) was at first merely astronomical, and may consequently reject it from our enquiry into the historical or civil chronology of India."

Since the time of Sir W. Jones, Bentley, in his "Astronomy," (page 26), as before stated, says that the astronomers in 945 B.C., among other things, had determined the rate of precession of the Equinoxes, which they found to be 3° 20′ in 247 Hindu tropical years and one month; this gives the precession = 48·56661″ or about 1·43″ too small.

For the purpose of examining the preceding construction, following backwards the order in which the Kalpa has been formed, we have :—

1 Kalpa	= 14 Manuwantaras+1 Krita.
The Manuwantara ..	= 71 Great Yugas+1 Krita.
A Great Yuga ..	= 10×432,000 years.
The Krita	= 4×432,000 years.
∴ The Manuwantara	= 710×432,000+4×432,000.
	= 714×432,000.

The Kalpa = (14×714+4) 432,000 = 4,320,000,000.

It is seen that this number consists of two factors, 14×714+4, which has the form $m.n+r$ = 10,000 and the co-efficient 432,000.

The form of the number shows that its inventors had an especial design in view in its construction, *i.e.*, to multiply the Kali period with the significant figures 432, unchanged. If they had no other design, there would have been no reason why they should have deviated from the rule laid down in the Institutes of Menu, which only required that they should multiply the divine age by a thousand. If they had merely wished to multiply 432,000 by 10,000, they would not have taken the trouble to have put the operation into such a singular form. It is clear that they did not wish to alter the factors already existing, in the Kali Age, namely,

$60 \times 60 \times 60 \times 2$, and that they especially wished to multiply by 10,000, so that their system would still be in conformity with that which was established in the Institutes of Menu and in the Vedas.

Now, there are a great many ways in which they might have multiplied by 10,000, and the fact that they selected this special form ($14 \times 714 + 4$) shows design. The number is one out of the set, $m\,n + r = 10{,}000$.

If we take m to be any number less than 100 which is not one of the eleven factors of $5^4 \times 2^4$ (each of which would divide 10,000 without a remainder), it would find by division a number which would have the form $m\,n + r$, and there would be 89 such cases, thus,

$$\begin{aligned} 10{,}000 &= 3 \times 3{,}333 + 1 = 6 \times 16{,}666 + 4 = 7 \times 1{,}428 + 4. \\ &= 9 \times 1{,}111 + 1 = 11 \times 909 + 1 = 12 \times 833 + 4. \\ &= 13 \times 769 + 3 = 14 \times 714 + 4. \end{aligned}$$

And so on, we might go through the whole of the 89 cases.

Out of all these cases, it is incredible that the particular form $14 \times 714 + 4$ should have been selected by chance.

Let us for a moment suppose that the astronomers who invented the Kalpa had made the discovery that in 714 years the Solstice (as a close approximation) had retrograded 10°, as, for example, from coincidence with Regulus to about 1° short of the beginning of the Nacshatra Magha, through which point the great circle, the line of the Rishis, before referred to, was assumed to pass. Then the Solstice would have gone back 14° in 1,000 years, or 140° in 10,000 years.

In 140° there are 504,000 seconds, which, divided by 10,000 gives the precession = 50·4″.

That this coincidence, out of so many adverse cases, could have happened by chance is not only improbable, but scarcely possible.

It is not surprising that the true nature of the Kalpa was not known to the later astronomers, who, in different ages, revised and condensed the various editions of the Siddhantas, when we consider the sacred and mysterious character given to it, concealed, as the precession was, amid a cloud of words, and guarded by the sacred names of the Gods, a circumstance which would seem to forbid the meddling of profane or common men. Moreover, the compilers of the Siddhantas and later Hindu astronomers may have considered it merely as a great number, which they had seldom occasion to use. But even if they could have had any suspicion of its true nature, and if its connection with the precession could have been supposed by them as probable, they would have been prevented from pursuing the investigation by the existence of rules, which gave a less perfect value of the precession than that of the Kalpa. One of the best of these is that contained in the Surya Siddhanta itself, which gives 54" for it. This rule was accepted by Bhaskara and other writers; it is referred to by the author of the Ayeen Akbery, by European writers in India, and also by Bailly and Playfair, as a remarkably close approximation for a period so early.

If this precession of 54" had been taken in the form of the number already referred to, it would have corresponded with $m = 15$, or $10{,}000 = 15 \times 666 + 10$, which is totally different in form from that of the Kalpa.

Again—$m\ n + r = 10{,}000$ is just one of the numerous indeterminate equations of the Hindus, which abound in their mathematical astronomy—equations the solution of which their early writers sometimes challenge each other in grotesque language.

The seventh and eighth sections of the XVIII. Chapter of the Brahmegupta Siddhanta are made up of astronomical questions of this nature, and a few specimens have been already given p. 169.

At the end of the chapter Brahmegupta extols a practice which would appear to have been prevalent in his time, of reciting such algebraical problems and proposing them for solution "in an assembly of the people."

HINDU ASTRONOMY.

PART II.

CHAPTER I.

DESCRIPTION OF THE SURYA SIDDHANTA.

One of the best known of the astronomical works of the Hindus, which has descended to the present time, written in the Sanscrit language, is the Surya Siddhanta. It is, however, a work adapted not so much for the schools as for the observer, and intended to instruct, not so much in the principles of the science as in the application of the rules.

The reader is directed to add and subtract, to multiply and to divide, and extract the square roots of the numbers he uses, and in the end he will find the result will agree with his observations.

The work itself is a compilation—a collection of aphorisms, a syllabus of formulæ—expressed in words so brief, and exact, as to become almost unintelligible, and requiring a great exercise of modern mathematical knowledge to discover their meaning, to test their accuracy, and to ascertain how far they apply to the subjects they refer to.

The first chapter (Par. 1—8) begins with an invocation, and this is followed by a short introduction regarding the origin of this work.

The science of astronomy is said to have been communicated by the sun to a great demon or spirit named Maya, through the agency of a man born from himself.

The figurative language here used might perhaps be translated to us, as the spirit or natural genius of Maya, by which men were led to the study of mathematics and astronomy, in which, by aid of the sun, information regarding the moon and the planets (whose paths were all near the Ecliptic) might be obtained. Observation showed that sometimes at points of the crossing of their paths, their meetings with each other occasioned partial extinction of their lights, as in eclipses, occultations, and other phenomena. The knowledge thus obtained was held to be a divine knowledge—a revelation not to be divulged to irreligious or common people, but only to disciples, who received it as a sacred and secret communication.

The ninth and tenth paragraphs make a distinction between time: First, as endless and continuous; and, secondly, as that which can be known.

The latter is of two kinds, one called Murta (measurable), the other Amurta (immeasurable).

(11) Time that is measurable is that which is in common use, of which a table of units is formed, beginning with the prana, which consists of four of our seconds. The pala contains six pranas.

(12) The ghatica is 60 palas, and the Nacshatra Ahoratra, or sidereal day and night, contains 60 ghaticas. A Nacshatra Masa, or sidereal month, consists of 30 sidereal days.

A distinction is made between the sidereal day and the Savana, or terrestrial day, the former being uniform, and the latter reckoned from one sunrise to the next, is, of course, variable, but, in the aggregate for the year, it is of the same length.

The Savana month consists of 30 Savana days.

(13) The Lunar month consists of 30 Lunar days or Tithis. It is the moon's synodic period from one new moon to the next, and the thirtieth part of this period is, therefore, the Lunar day.

A Solar, or Saura year consists of 12 Saura months, and this is called a Deva day, or a day of the Gods.

The Saura month is the time which the sun takes to move from the beginning of one sign of the Zodiac to the next, and the Solar or Saura year is reckoned to begin at the Sancranti of Mesha, that is, at the moment when the sun enters that sign. This is also the first point of Aswini, and within a few minutes in arc of the star Revati, identified, according to Colebrooke (Essays, Vol. II. p. 464), as the star ζ Piscium, whose longitude, rectified to the beginning of 1800 A.D., was 17° 4′ 48″, and latitude 13′ 11″ S., so that, reckoning the mean annual precession at 50″, the longitude of the first point of Aswini is 18° 24′ 48″ from the Vernal Equinox of the present year 1896.

From paragraph (13), it would appear that the Saura year, or Deva day, when first instituted, consisted of 360 Saura days, each of which was considered to be the time taken by the sun to move over a degree of the Ecliptic. Consequently, these astronomical days were of unequal length, the mean of which would be greater than the true mean Solar day, since the mean velocity each day is only 59′ 0·7″.

That this ancient day is still retained in the Siddhantas implies that the mean Solar day of the later works on astronomy has not the same meaning as the Saura day of ancient times.

From the 15th to the 19th paragraphs, we have the formation of the large periods of time the Maha-Yuga and the Kalpa, the rules of which have been already cited and explained, and which were intended to fix, in the past, certain epochs, at each of which, for different purposes, it was found convenient to assume a common origin for astronomical calculations, and from which epoch, in order to simplify the computations, the sun, the moon, and the planets, with their nodes and apsides, might be assumed to start together from the same point of the Ecliptic, namely, the first point of Aswini.

The position or place of each planet for any given time after the epoch would then depend only on determining, by simple proportion, between the number of revolutions in a Maha-Yuga, the corresponding required revolutions in a given time from the epoch.

Paragraph (20) assumes the Kalpa to be a day of Brahma, and his night to be equal to his day.

In 21 it is stated that the age of Brahma consists of 100 Kalpas, one-half of which has passed away, and that the present Kalpa is the first of the remaining half.

(22-23) From the beginning of the present Kalpa there have passed away six Manus with their Sandhis; and the Sandhi which is at the beginning of the Kalpa, 27 Maha-Yugas, and the Krita-Yuga at the beginning of the 28th.

(24) The sum of these is 5,474,400 Deva years, from which is to be subtracted 47,400 Deva years, which were passed by Brahma in creating animate and inanimate things. The remainder is the time elapsed from the beginning of the present order of things before the end of the Krita-Yuga—5,427,000 Deva years.

In paragraph 25 it is stated that, "the planets in their orbits go rapidly and continually with the stars towards the West, and hang down at an equal distance as if overpowered (or over-matched in speed) by the stars."

(26) "Therefore the motions of the planets appear towards the East, and their daily motions, determined by their revolutions, are unequal to each other in consequence of the circumferences of their orbits; and by this unequal motion they pass the signs."

(27) "The planet which moves rapidly requires a short time to pass the signs, and the planet that moves slowly passes the signs in a long time."

In explanation of these three passages it is to be understood that Hindu astronomers hold the opinion that the planets move in their orbits with the same actual linear velocities, and that it is owing to the circumference of the orbits being of greater or less dimensions that the planets moving in them appear to move more slowly or more rapidly.

Thus, the Hindus found that the circumference of the moon's orbit was 324,000 Yojanas, and the periodic time being 27.3216 days, the daily motion in her orbit would be 11,858¾ Yojanas nearly. Then, according to this theory, for any other planet the circumference of the orbit $= P \times 11,858\frac{3}{4}$ Yojanas; P being here put for the periodic time of the planet.

This opinion that the motion of all the planets was caused by a velocity in their orbits, which was the same for all alike, was prevalent not only in the East, but also in Europe even to the times of Kepler and Newton. This is evident from the manner in which Kepler combatted this doctrine, and the important use he made of it. Soon after the death of Tycho, Kepler "made many discoveries from Tycho's observations. He found that astronomers had erred from the first rise of the science in ascribing always circular orbits and uniform motions to the planets. . . . . He easily saw that the higher planets not only moved in greater circles, but also more slowly than the nearer ones, so that, on a double account, their periodic times were greater. Saturn, for example, revolves at a distance from the sun nine times and a half greater than that of the earth's, and the circle described by Saturn is in the same proportion; and, as the earth revolves in a year, so, if their velocities were equal, Saturn ought to revolve in nine years, but not in so great a proportion as the squares of those distances (the square of 9½ being 90¼) for if that were the law of their motions, the periodic time of Saturn ought to have been above 90 years.

"A mean proportion between that of the distances of the planets, and that of the squares of those distances, is the true proportion of the periodic times, as the mean between 9½ and its square, 90¼, gives the periodic time of Saturn in years.

"Kepler, after having committed several mistakes in determining this analogy, hit upon it at last in 1618 (May 15th), for he is so exact as to mention the precise day when he found that 'the squares of the periodic times were always as the cubes of their mean distances from the sun.' "—(Maclaurin).

(28) Gives the table of circular units, which, with Indian names, is the same as that in general use, beginning with the Vikala, or second of arc, thus:—

60 Vikalas	make	1 Kala, a minute.
60 Kalas	„	1 Ansa, a degree.
30 Ansas	„	1 Rasi, a sign.
12 Rasis	„	1 Bhagana, a revolution.

From verses (28) to (33) there are given in detail the number of revolutions made by each of the planets, with the nodes and apogees of the moon, in a Maha-Yuga; they are given in detail, but are here arranged in Table I.

They must have been derived from similar numbers previously existing, before the time when the Surya Siddhanta was compiled, in forms which had undergone numerous alterations and corrections, from time to time, to bring them into agreement with later observations.

This table is employed in all the exact problems of Indian astronomy.

It is assumed that, at the Creation, the sun, the moon, and the planets, with their apsides and nodes, began their motions together from nearly the same first Meridian, and at the beginning of each Maha-Yuga the sun, the moon, with the moon's apsides and nodes, were reckoned to be then in conjunction in the line joining the first point of Aswini, or of Mesha, with the centre of the earth.

Hence, the mean places of the planets, and most of the problems relating to longitudes, such as those of conjunctions and oppositions were determined by rules which depended only on simple proportion, when the epoch was a given date in the past.

TABLE I.

Revolutions of the Planets, &c., in a great Yuga:—

The Sun	4,320,000
The Moon, Sidereal revolutions ...	57,753,336
Mercury	17,937,060
Venus	7,022,376
Mars	2,296,832
Jupiter...	364,220
Saturn	146,568
Moon's Synodic revolutions	53,433,336
„ Apogee...	488,203
„ Nodes	232,238
The Number of Savana days in a Maha-Yuga is	1,577,917,828
The Number of Lunar days in a Maha-Yuga is	1,603,000,080

From verse (34) the number of sidereal revolutions in a Great Yuga is 1,582,237,828. The number of risings of a planet in a Great Yuga is the difference between the number of sidereal revolutions and the planet's own revolutions.

(35) The number of lunar months is equal to the difference between the revolutions of the moon and those of the sun.

The number of Adhimasas, or additive months, is the difference between the lunar months and the solar.

(36) The difference between the lunar days and the savan days is the number of subtractive days.

(37) There are 1,577,917,828 terrestrial or savan days, and 1,603,000,080 lunar days in a Great Yuga.

(38) Also 1,593,336 additive months, and 25,082,252 subtractive days, 51,840,000 solar months in a Great Yuga.

The large numbers given in verses (34) to (39) are of great importance in the construction of the Hindu luni-solar year.

If the ratio be formed between the additive months and the solar months in a Yuga, we have

$$\frac{\text{Additive Months}}{\text{Solar Months}} = \frac{1,593,336}{51,840,000} = \frac{1}{32 \cdot 53603}$$

In like manner, the ratio of the additive months to the lunar months in a Great Yuga is

$$\frac{\text{Additive Months}}{\text{Lunar Months}} = \frac{1,593,336}{53,433,336} = \frac{1}{33 \cdot 5355}$$

Hence, the ratio of the solar months to the lunar months:

$$\frac{\text{Solar Months}}{\text{Lunar Months}} = \frac{32 \cdot 53603}{33 \cdot 5355}$$

Which shows that for the intercalation, one month is to be added to 32½ solar months in order to find the corresponding number of lunar months.

The double month, called Adhimasa, thus intercalated, makes a month of 60 lunar days or Tithis.

Another adjustment is also required in the luni-solar year, for the difference between the lunar and solar day.

The lunar day is the time which the moon takes in separating from the sun, to the extent of 12° of arc, and this is the 30th part of the moon's synodic period of 29.53058 days.

The solar month, understood above, is the 12th part of the solar year of 365.25875 days.

Now, if at any time the beginning of the lunar day was coincident with that of the solar day, being a shorter day, it would terminate sooner than the solar day, and the difference would increase daily, and the time when they would begin together again could be determined from the above.

But this is effected by means of what are called the subtracted, or omitted, days.

First, if the ratio be formed of the number of subtracted days and the savan or terrestrial days in a Maha-Yuga, we have

$$\frac{\text{Subtractive Days}}{\text{Savan Days}} = \frac{25,082,252}{1,577,917,828} = \frac{1}{62 \cdot 9097}$$

Again, the ratio being formed of the subtractive days and the lunar days in a Maha-Yuga

$$\frac{\text{Subtractive Days}}{\text{Lunar Days}} = \frac{25,082,252}{1,603,000,080} = \frac{1}{63\cdot9097}$$

$$\text{Or the ratio } \frac{\text{Solar Days}}{\text{Lunar Days}} = \frac{62\cdot9097}{63\cdot9097}$$

The correction for the two kinds of days is made by subtracting one day from 63.9097 lunar days, in order to find the corresponding number of savan or solar days.

Bhaskara, following Brahmegupta, makes the Avama, or subtractive day, to occur in $64\frac{1}{11}$ tithis or lunar days, the Avama being a savan day. The slight difference in the calculation may be owing to Brahmegupta's numbers being somewhat different from those of the Surya Siddhanta.

The effect produced by the added month upon the calendar is to put back the names of the lunar months, and to change the times of the holidays and festivals. Those occurring in the double month of 60 days are retained in their own proper months; but those which follow will all be advanced, with their respective months, an entire lunation.

Paragraphs 41 to 44 give revolutions of the Apogees and of the Nodes in a Kalpa:—

	Apogees.	Nodes.
Of the Sun	387 ...	—
„ Mercury	368 ...	488
„ Venus	535 ...	903
„ Mars	204 ...	214
„ Jupiter	900 ...	174
„ Saturn	39 ...	662

Indian astronomers were not universally in agreement regarding a suitable epoch.

The commentator of the Suryá Siddhanta, when remarking upon the difference of opinion that existed as to whether the computations ought to begin with the Kalpa, or with the period stated in the Surya Siddhanta, says:—"It is of no consequence to us which, since our object is only to know which period answers for computation of the planetary places in our time, not at the beginning of the Kalpa. The difference found in computing, according to Brahmegupta and the Munis, must be corrected by an allowance of Bija, or by taking that difference as the Kshepa; but the books of the Munis must not be altered, and the rules given by Brahmegupta, Varah-Acharija, and Aryabhatta may be used with such precautions. Any person may compose a set of rules for the common purposes of astronomy, but with regard to the duties necessary in eclipses, the computation must be made by the books of the Munis, and the Bija applied."

The date of the epoch, then being given, at which time the planet is supposed to be at the first point of Aswini, and the number of days since that time being consequently known, we have the following proportion:—

Days in a Maha-Yuga : Revolutions of Planet in Maha-Yuga
: : Elapsed days : Revolutions in elapsed time, or

$$\left.\begin{array}{l}\text{Revolutions in the}\\ \text{elapsed time from}\\ \text{the Epoch}\end{array}\right\} = \frac{\text{Revolutions in Maha-Yuga}}{\text{Days in Maha-Yuga}} \times \left\{\begin{array}{l}\text{Elapsed days}\\ \text{since Epoch}\end{array}\right.$$

The result will in general be found to consist of an integral number of revolutions, and a fraction; rejecting the integer, the fraction, if any, will be the mean longitude of the planet from the first of Aswini, on the first Meridian, namely, that of Ujjaini or on the Meridian at Lanca.

To find the Arghana, *i.e.*, the number of civil days elapsed from the beginning of Creation, when all the celestial moveable bodies—the sun, the moon, and planets, with their nodes and apsides—were in conjunction, up to the present day.

(48-51) The elapsed years from the creation to the end of the Satya-Yuga, is reckoned to be ... ... 1,953,720,000 solar years

Satya-Yuga, is reckoned to be ...	...	1,953,720,000	solar years
Years of the Trita and Dwapara	...	2,160,000	"
Time elapsed from the creation to the beginning of the Kali-Yuga	...	1,955,880,000	"

To this great number is now to be added the years that have transpired since the beginning of the Kali-Yuga to the initial day of the present year, which, for 1895 A.D., is the 12th of April, the beginning of the Hindu solar year, the day on which the sun enters the Hindu Zodiac, in the first point of Mesha, or, rather, the first point of the Nacshatra Aswini.

The present Kali-Yuga is estimated by Bailly and others to have begun at midnight between the 18th and 19th February, 3102 B.C., the sun being then on the Meridian of Lanca; the elapsed time from the Creation, therefore, according to the Hindu account, would be—

From the creation to the Kali-Yuga	1,955,880,000	years
And from the Kali-Yuga to the year 1895 A.D.	3,101 +1895	" "
Or the elapsed solar years from the Creation to April 12th, 1895, *i.e.*, to the mean Sancranti, when the sun enters the sign of Mesha ...	1,955,884,996	"
This number when multiplied by 12 gives the elapsed solar months up to the Mesha Sancranti	23,470,619,952	

Now, as the day on which the elapsed time is required, may be any day in any month after the beginning of the year, as, for example, the day on which an eclipse will happen, suppose the months and days to be m and d respectively, d being lunar days of the current lunar month; the rule proceeds to add the number of months m to the above elapsed solar months, and the elapsed solar months (say E S M) nearest to the given time will then be $23,470,619,952+m$. To make these solar months lunar, the addi-

o 2

tive months proportional to E S M must be computed and added to E S M.

To abbreviate the calculation assuming initial letters for the terms, let A M be the additive months in a Yuga, and S M be the solar months in a Yuga. Then the elapsed additive months will be

$$E\,A\,M=\frac{A\,M}{S\,M}\times E\,S\,M,$$

adding this to E S M we have the elapsed lunar months nearest to the given day, or to the end of the last lunar month, *i.e.,*

$$E\,L\,M=\left(1+\frac{A\,M}{S\,M}\right)E\,S\,M$$

This number of elapsed lunar months, multiplied by 30, and increased by *d*, is the number of elapsed lunar days, or

$$E\,L\,D=30\left(1+\frac{A\,M}{S\,M}\right)\times E\,S\,M+d \text{ tithis.}$$

The rule now requires that these lunar days, or tithis, should be converted into civil days, for which purpose the elapsed subtractive days are to be computed.

Again, to abbreviate, let O D be the omitted days to be subtracted in a Yuga, and L D the lunar days in a Yuga. Then the elapsed subtractive days will be, by proportion,

$$E\,O\,D=\frac{O\,D}{L\,D}\times E\,L\,D$$

The elapsed omitted days being subtracted from the elapsed lunar days, will give the elapsed civil days to the end of the last lunar day; hence,

the elapsed civil days $=\left(1-\frac{O\,D}{L\,D}\right)\times E\,L\,D$ from the creation

$$=\left(1-\frac{O\,D}{L\,D}\right)\left\{30\left(1+\frac{A\,M}{S\,M}\right)\times E\,S\,M+d\right\}$$

In this formula E S M = 23,470,619,952+*m*.

S M = 51,840,000 solar months.

L D = 1,603,000,080 lunar days or tithis.

A M = 1,593,336 additive months.

O D = 25,082,252 subtractive days.

(51) The rulers of the days of the week are then found and are indicated by the remainders on dividing the elapsed civil days by 7; thus, if there is a remainder 1, it indicates Ravi-Var, or Sunday the lord of which is Ravi, or the sun; the remainder 2 indicates Soma-Var, the lord being Soma, or the moon; remainders 3, 4, 5, 6, indicate Mangula-Var, Budha-Var, Vrihaspati-Var, Sucra-Var, and Sani-Var, the lords of which are Mars, Mercury, Jupiter, Venus and Saturn.

(52) Rules are also given for finding the lords of the month and of the year.

Rule (53) gives the method of finding the mean place of a planet at any time, referred to Lanka, the first Meridian in India. The number of the elapsed days from the epoch is to be multiplied by the number of revolutions of the planet in a Yuga, and the product divided by the number of terrestrial days in a Yuga; the quotient will be in general a complete number of revolutions, with a remainder. Of these, the revolutions are to be rejected and the remainder only retained; this is to be reduced to signs, degrees, minutes, etc., and in this form it will be the mean place of the planet, or its longitude from Aswini.

(54) In the same way, the mean place of the apogee of the planet is to be found, and the same calculation applies to the nodes; but the nodes, having a retrograde motion, the result in signs, degrees, etc., must be subtracted from 12 signs.

In Rule (55) we have the direction for finding the so-called present Samvatsara.

A Samvatsara is, as before mentioned, the time which Jupiter takes, by his mean motion, to move over each sign of 30°, and which nearly corresponds with one of our solar years. Jupiter's cycle consists of 60 such periods, each of which has a name, that of the first in the cycle being Vijaya.

The rule directs that the number of elapsed revolutions of Jupiter is to be multiplied by 12, and to the product is to be added the number of signs intervening between the place occupied by Jupiter and the beginning of Stellar Mesha; the sum is then to be divided by 60, which will consist of an integral quotient, and a remainder; the remainder, reckoned from the period called Vijaya, is the required Samvatsara.

Rule (56) suggests the beginning of the Trita-Yuga, as a convenient epoch from which to compute the elapsed time, for the purpose of finding the mean places of the planets.

Rule (57) continues that at this epoch the mean places of the planets, with the exception of their apogees and nodes, were together coincident in the first point of Mesha.

Rule (58) further states that the place of the moon's apogee was then 9 signs, and her ascending node 6 signs, and that the apogees and nodes of the five planets had some uncertain amount of signs and degrees.

In (59) it is said that the diameter of the earth is 1,600 Yojanas, and that the product of the diameter by the root of 10 will be the circumference. This method of finding the circumference of a circle from its diameter is only one of many that were employed for this purpose in the Siddhantas, some of them giving much nearer approximations to the true ratio of the circumference to the diameter. Amongst them are $\frac{4966}{1581}$, $\frac{22}{7}$, and the still nearer value $\frac{119208}{3794}$ which is given by Bhascara in the Siddhanta Siromani. It is said also that Aryabhatta gives $\frac{62832}{20000}$, and that it was only for convenience of calculation that the circumference was taken as diameter $\times \sqrt{10}$.

The process by which the Hindus obtained an approximate knowledge of the circumference and diameter of the earth, was, no doubt, with the aid of mid-day shadows, cast on planes by lofty objects, such as the spire of a temple, or vertical poles in different places on the same Meridian at the same time of the year (for instance, an

Equinox or a Solstice), the distance between the places being known or measured.

The Aryans, in their migrations and progress Southward, must have observed that the shadow of the same object, as, for example, a tent-pole (an observation necessary for their religious observances at the time of the Solstice), would continually diminish as the latitudes of the places arrived at decreased.

On account of the great distance of the sun, rays of light from it may be considered as coming to all parts of the terrestrial hemisphere on which they fall, in parallel lines.

For, if we suppose two places A and B are on the same Meridian, the earth being assumed as a sphere, rays to the summits of two vertical objects, A M and B N, would be in parallel lines, S M and

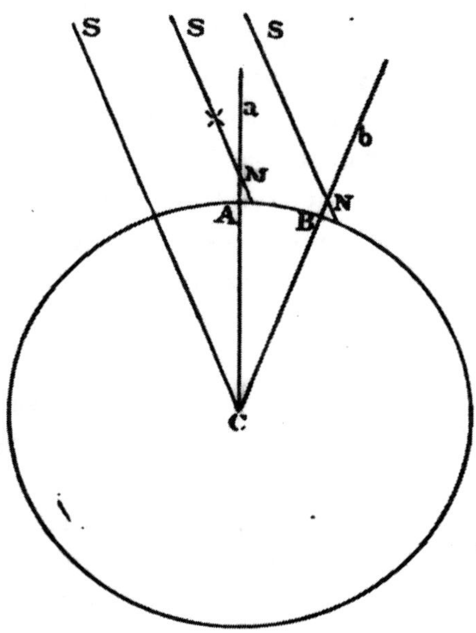

S N forming the vertical angles S M *a* and S N *b*, or the Zenith distances of the sun at the two places. If C be supposed to be the centre of the earth, and S C the direction to it, of a ray from the

sun, A C and B C being radii from A and B to the centre, then the angle A C S is equal to the angle *a* M S, and the angle B C S is equal to the angle *b* N S. Therefore, the angles A C S and B C S are the zenith distances of the sun observed at A and B at the same time.

Now, the angle A C B is the difference between these two angles, and it is measured by the arc A B, the difference of latitude between the two places. Hence, if the arc A B (the distance between A and B) be measured in Yojanas, and assumed equal to *a*, then, by proportion—

$$\text{A B in degrees} : 360^\circ : : a : \text{earth's circumference}$$
$$\text{or the earth's circumference} = \frac{360^\circ}{\text{A B}^\circ} \times a \text{ yojanas.}$$

The method of measuring the arc of a degree of the Meridian, the earth being considered as a sphere, was well known to the Hindu astronomers, as is clearly shown by Bhascara in the Siddhanta Siromani, in his refutation of the opinion of the Jains, who adopt the doctrines of the Puranas, and, among other fanciful theories, describe the earth as a great level plane. He says:—"As the earth is a large body, and a man is exceedingly small, the whole visible portion of the earth consequently appears to a man on its surface to be perfectly plane.

"That the correct dimensions of the circumference of the earth have been stated may be proved by the simple rule of proportion (he gives $1581\frac{1}{24}$ Yojanas as the diameter and 4967 Yojanas as the circumference in this mode: Ascertain the difference in Yojanas between two towns in an exact North and South line, and ascertain also the difference of the latitudes of those towns; then say if the difference of latitude gives this distance in Yojanas, what will the whole circumference of 360° give?

"As it is ascertained by calculation that the city of Ujjayni is situated at a distance from the Equator equal to the one-sixteenth part of the whole circumference, this distance, therefore, multiplied by 16, will be the measure of the earth's circumference. What reason, then, is there for attributing such (50,000 Yojanas) an immense magnitude to the earth?

"For the position of the moon's cusps, the conjunctions of the planets, eclipses, the times of risings and settings of the planets, the lengths of the shadows of the gnomon, etc., are all consistent with this (estimate of the extent of the) circumference, and not with any other; therefore, it is declared that the correctness of the afore-measurement of the earth is proved, both directly and indirectly (directly by its agreeing with the phenomena, and indirectly by no other estimate agreeing with the phenomena)."

The Siddhantas did not all agree regarding the numerical dimensions of the diameter and circumference.

ON THE METHOD OF FINDING THE LONGITUDE.

In the geometry of the sphere, the small circle, or parallel of latitude, of any place on the earth's surface, is referred to in Indian astronomy as the rectified circumference (the Sphuta), and Rule (60) gives the ordinary method by which it is determined, thus:

$$\left.\begin{array}{l}\text{The rectified}\\\text{circumference}\end{array}\right\} = \frac{\text{Earth's Circumference}}{\text{Radius}} \times \left\{\begin{array}{l}\text{Sin Colatitude of}\\\text{place.}\end{array}\right.$$

The same rule gives a correction to be applied to the mean place of a planet, calculated for midnight on meridian of Lanca, to make it serve for a place that may be East or West of that Meridian, the so-called middle-line, or Madhya-Rekha.

This correction is called the Desantara Correction, and its amount is found from

$$\text{Desantara} = \frac{\text{Distance in Yojanas from mid-line}}{\text{Rectified Circumference}} \times \left\{\begin{array}{l}\text{Planets daily}\\\text{motion in minutes}\end{array}\right.$$

This correction is applied also by some astronomers to the place of a planet computed for sunrise.

Rule (61) directs the Desantara to be subtracted from the mean place of the planet at midnight on the first Meridian, if the given place of the observer be East of the mid-line; but if it be West, it is to be added to the computed place of the planet.

Rule (62) states that the cities of Ujjayni, Rohitaka, and Kurukshetra are all on this mid-line of the earth; and the line is also assumed to pass through the hypothetical place Lanca, a place supposed to be on the Equator, but which would necessarily be a point of the Indian Ocean about 6° South of Ceylon.

(63, 64, 65) These rules give the method of finding the longitude of a given place on the earth, from observations of the beginning or ending of the total darkness in a lunar eclipse.

If in the eclipse, seen at the place of the observer, the total darkness begins or ends after the instant for which it has been computed to begin or end at the middle line, then the place of the observer is East of the middle line; but if the beginning or ending be before the computed time, the observer's place is West of the middle line.

Next, the difference is to be found between the observed time at the place and the computed time on the middle line. This difference is called the Desantara Ghatikas.

Then the distance of the place of the observer from the middle line in Yojanas

$$= \frac{\text{the Desantara Ghatikas}}{60} \times \text{the Rectified Circumference}$$

From this distance in Yojanas the minutes of the Desantara are to be found and applied to the places of the sun and moon (by means of Rules (60) (61).

The mean places of the planets, determined by preceding rules, are for the midnight of a given day. Rule (67) supposes that the mean places may be required for a time in Ghatikas, before or after midnight, on the day for which the places have been computed, and then the corrected place would be

$$= \text{Computed place at midnight} \pm \frac{\text{Time in Ghatikas}}{60} \times \left\{ \begin{matrix} \text{Planets daily} \\ \text{motion} \end{matrix} \right.$$

The inclination of the moon's path to the Ecliptic was supposed to be a deflection caused by the node. And in (68), the greatest deflection is stated to be 4° 30′.

This, however, is under the modern estimate of 5° 9′, but the inclination is variable, the greatest inequality being 8′ 47″.

(69-70) In like manner, the orbits of the planets were considered to be deflected respectively by their nodes, the inclinations, or mean greatest latitudes, being put as under:—

The Moon	...	...	270′ or 4° 30′
Mars ...	...	...	90′ or 1° 30′
Mercury	...	...	120′ or 2°
Jupiter	...	...	60′ or 1°
Venus ...	...	...	120′ or 2°
Saturn ...	...	...	120′ or 2°

In the subjoined table, the mean sidereal periodic times of the planets have been computed in mean solar days, by dividing the number of Savan days in a Maha-Yuga, or 1,577,917,828, by the number of revolutions of each of the celestial bodies given in Table I.; and they are compared with the corresponding periodic times taken from Herschel's Astronomy, the latter being arranged in the third column of the table.

Mean sidereal periods compared with those given in Herschel's Astronomy:--

	From Surya Siddhanta Mean Sidereal Periods in mean Solar days.	*From Herschel's Astronomy* Mean Sidereal Periods in mean Solar days.
The Earth	365·25875	365·2563612
The Moon	27·32167	27·32166148
The Moon's Nodes...	6794·443	6793·39108
The Moon's Apogee	3231·2	3232·575343
Mercury	87·9697	87·96925
Venus	224·69792	224·7007869
Mars	686·9975	686·9796458
Jupiter	4332·3206	4332·5848212
Saturn	10765·773	10759·2198174

For a similar purpose another table has also been computed from Table I., showing the synodic periods of the planets, as compared with corresponding periods taken from Woodhouse's Astronomy.

These have been obtained from the form

$$\text{Synodic period} = \frac{\text{Solar days in a Great Yuga}}{\text{Revolutions of Planet}-\text{revolutions of Sun}}$$

Thus, in the case of the moon:—

$$\text{The Moon's Synodic period} = \frac{1{,}577{,}917{,}828}{5{,}753{,}336-4{,}320{,}000}$$

$$= 29{\cdot}530586 \text{ mean solar days}$$

In the same way the other synodic periods in the table have been found.

	Synodic period in Mean solar days, Surya Siddhanta.	Woodhouse's Astronomy.
The Moon	29·530586	29·530588
Mercury	115·88	115·877
Venus	583·9	583·92
Mars	779·924	779·936
Jupiter	398·89	398·867
Saturn	378·08	378·09

The method of computing these periods gives thus a close comparison with that employed in our Astronomy.

ON THE MOON'S HORIZONTAL PARALLAX, AND THE DISTANCE OF THE MOON FROM THE CENTRE OF THE EARTH.

In the Surya Siddhanta we have the moon's horizontal parallax given in the form:

$$\frac{5059}{324000} \times 3438 \text{ minutes of arc which redneed} = 53{\cdot}681'$$

This as a mean is smaller than it ought to be, for the Horizontal Parallax.

At the greatest — 61·533′

At the least — 52·88′

But, taking into consideration that the effects of refraction were not known in India, nor even in Europe till the time of Tycho and Kepler (the latter of whom gave the first treatise on refraction) it may be conceded that the Indian horizontal parallax of the moon was a fair approximation.

Supposing, then, that an equally close approximation of the diameter of the earth had been obtained, the moon's distance would have been fairly represented by 51,566 Yojanas, on the supposition that the diameter of the earth was 1,610 Yojanas; and this would make the circumference of the moon's orbit 324,000 Yojanas.

This was the number assumed in all relative calculations.

As the parallax of the sun could not be obtained by direct observations, the Indian astronomers had recourse to the theory referred to under verse (27) Part II, Chap. I., by which they supposed that all the planets moved in the respective orbits with the same actual linear velocity. By this hypothesis they accounted for the apparent slowness of some of them by their having to travel over orbits of greater diameter, and the circumferences were supposed to vary directly as the periodic times.

Taking, then, the circumference of the moon's orbit, or 324,000 Yojanas, and its periodic time, 27,176 days, as constants, they computed the circumferences of the orbits of the sun and planets, etc., by simple proportion, from a form equivalent to

$$\left.\begin{array}{l}\text{Circumference of}\\ \text{orbit in Yojanas}\end{array}\right\} = \frac{324{,}000}{27{\cdot}176} \times P, \text{ where } P = \text{periodic time of planet.}$$

In this way, the supposed circumferences of the orbits were found and given, as under, in Chapter XII. of the Surya Siddhanta :—

The Moon's Orbit...	324,000 Yojanas.
Sighrocheha (Apogee) of Mercury	1,042,000 „
„ „ Venus...	2,664,637 „
Orbits of Sun, Mercury and Venus	4,331,500 „
„ Mars	8,146,909 „

Orbits of Moon's Apogee... ...	38,328,484	Yojanas.
" Jupiter	51,375,764	"
" Saturn	127,668,255	"
" Fixed Stars	259,890,012	"
The circumference of the Brahmandee, the egg of Brahma, in which the sun's rays are spread	18,712,080,864,000,000	"

CHAPTER II.

ON THE METHOD EMPLOYED BY THE HINDUS FOR FINDING THE TRUE PLACE OF A PLANET FROM ITS MEAN PLACE.

It was well known to the Hindus that a supposed uniform motion in a circle about the earth did not really represent the true motion of a planet in its orbit, although the hypothesis served sufficiently to determine the mean motions and the mean place of a planet when deduced from observations carried on for lengthened periods. They knew that every planet in its course was subject to great irregularities, the motion undergoing continual changes. At one time it would be direct towards the East, until the planet reached a stationary point, where it would seem to be at rest; then a retrograde motion would begin, and continue for a time, till another stationary position was reached, and the Eastward motion would be repeated.

It was to account for these irregularities that the Epicycle was invented.

By the Greeks this contrivance was ascribed to Apollonius. He conceived that a planet in its course described, with uniform motion, the circumference of a circle, called the Epicycle, whose centre moved uniformly in the circumference of another circle, called the deferent, the centre of which was the centre of the earth.

It was also supposed that, whilst the centre of the Epicycle was moving Eastward in the direction of the signs, the planet itself was moving in a direction contrary to that of the signs. By this hypothesis it was easy to show the various changes in the motions of the planets. This theory was generally adopted by Western nations, with the addition of other Epicycles, introduced by Ptolemy, as necessary for expressing the apparent motions with accuracy.

The Hindus had two methods for calculating "the true place" of a planet from its mean place, as determined by the rules of the Surya Siddhanta.

One of these methods resembled that of Apollonius, with this difference: that, whilst the planet moved uniformly in its Epicycle, whose centre moved in the deferent concentric with the earth, the Epicycle itself was conceived to be variable, the circumference being greatest when the planet was in an apsis (at Apogee or Perigee, the "true" and mean places being then coincident), and least when the planet was at a distance of 90° from those points.

The other method supposes that, while the mean place of a planet is a point moving uniformly Eastward, round the circumference of a circle whose centre is the earth, the planet also moves uniformly Eastward, in the same time, round the circumference of an equal, but eccentric, circle, whose centre is situated in the line joining the Apogee with the centre of the earth, the distance from it being the eccentricity.

These two methods of calculation, whether by assuming the motions as being in an eccentric or in an Epicycle, give exactly the same results; but it will be observed that, whereas the planet on the former hypothesis is conceived to move in the direction of the signs, on the latter hypothesis the apparent motion of the planet in the Epicycle would be in a contrary direction.

It is on this seeming inconsistency that Bhascara, in the Siromani, makes the following sensible remarks:—

"As the actual motion in both cases is the same, while the appearances are thus diametrically opposed, it must be admitted, therefore, that these expedients are the mere inventions of wise astronomers, to ascertain the amount of equation."

It is to the greatest equation of the sun's centre that Laplace refers when, after stating that the epoch of the Kali-Yuga was determined by calculation and not by observation, he says:—"But it must be owned that some elements of the Indian astronomy seem to indicate that they have been determined before this epoch

(3102 B.C.). Thus, the equation of the centre of the sun, which they fix at 2·4173°, could not have been of that magnitude but at the year 4300 before the Christian Era."

Preparatory to calculating the "true" places of the planets from their mean places, and for general purposes, a table was constructed of the sines and versed sines of the arcs of a circle.

These functions differ so remarkably from those which were in use in the works of Western nations, that they have, as it were, stamped upon them the genuine character of original productions.

We are so accustomed to the use of sines of angles, and sines of arcs to radius unity, that we are apt to believe that nothing else could be so simple; and yet a still more simple system was in use among the Hindus at times earlier in their history than that of the compilation of the Surya Siddhanta.

The peculiarity connected with these sines is this, that the radius of the circle from which the sines are calculated is really the so-called analytical unit, which, though suggested for use in modern times*, like a strange coin, never, for obvious reasons, obtained much currency.

As an angular unit it may thus be defined:—

If, in the circumference of any circle, an arc be taken equal in length to the radius of the same circle, the angle measured by this arc, or the degrees, minutes and seconds which it contains, will be 57° 17′ 44″ 48‴, etc.

This angle or arc reduced to minutes = 3437·746′.

The nearest whole number being 3438.

This number (3438) the Indian astronomers used as the radius, in calculating their table of sines, and whenever, in the rules given as solutions to their problems, in Spherical Trigonometry, they make use of the word "radius," this number is understood.

* By Dr. Hutton, Phil. Trans., 1783.

The table given is not very extensive, being only for arcs, multiples of $\frac{1}{8}$ of an arc of 30°, *i.e.*, of 3° 45′.

When sines of other arcs were required, they were found by proportional differences to the nearest minute in whole numbers (the use of decimals being unknown), and the approximation would correspond to about four places of decimals of our tables.

Two rules are given in the Surya Siddhanta for calculating a table of sines. Beginning with an assumed first sine, they proceed by progressive equal arcs, computing the succeeding sines, in order, from those previously found.

These rules are:—

(15) "The eighth part of the number of minutes contained in a sign (*i.e.*, of 30° or 1800′) is the first sine. Divide the first sine by itself, substract the quotient from that sine, and add the remainder to that sine; the sum will be the second sine."

(16) "In the same manner, divide successively the sines (found) by the first sine; subtract the (sum of) the quotients from the divisor, and add the remainder to the sine last found, and the sum will be the next sine. Thus, you will get twenty-four sines (in the quadrant of a circle whose radius is 3438)."

These rules express in words the operations implied in the formula

$$\text{Sin } (n+1) \text{ A} = \sin n \text{ A} + \sin \text{A} - \frac{\sin \text{A} + \sin 2 \text{A} + \text{etc.} \ldots . \sin n \text{ A}}{\sin \text{A}}$$

where sin A is the first sine.

For making $n=1$ and sin A=225′ or $\frac{1800'}{8}$

$$\text{Sin } 2 \text{ A} = \sin \text{A} + \sin \text{A} - \frac{\sin \text{A}}{\sin \text{A}} =$$

$$225+225-\frac{225}{225}=449, \text{ making } n=2.$$

$$\text{Sin } 3 \text{ A} = \sin 2 \text{ A} + \sin \text{A} - \frac{\sin \text{A} + \sin 2 \text{A}}{\sin \text{A}} =$$

$$449+225-\frac{225+449}{225}=671, \text{ making } n=3.$$

$$\text{Sin } 4 \text{ A} = \sin 3 \text{ A} + \sin \text{A} - \frac{\sin \text{A} + \sin 2 \text{A} + \sin 3 \text{A}}{\sin \text{A}} =$$

$$671+225-\frac{674+671}{225}=890, \text{ making } n=4.$$

$$\text{Sin } 5\text{ A}=\sin 4\text{ A}+\sin\text{ A}-\frac{\sin\text{ A}+\sin 2\text{ A}+\sin 3\text{ A}+\sin 4\text{ A}}{\sin\text{ A}}=$$

$$890+225-\frac{1345+890}{225}=1105.$$

and so on for $n=5$, 6, etc., throughout.

This formula may have been derived from the ordinary formula

$$\text{Sin }(\text{A}+\text{B})+\sin(\text{A}-\text{B})=2\sin\text{ A}\cos\text{ B},$$

which, with

$$\text{Sin }(\text{A}+\text{B})-\sin(\text{A}-\text{B})=2\cos\text{ A}\sin\text{ B},$$

were known to the Hindus, before the date of the Siromani, for Bhaskara says they were called Jaya-Bhavana, and that they were prescribed for ascertaining other sines.

Or the rule may have been derived from the formula

$$\text{Sin }(n+1)\text{ A}+\sin(n-1)\text{ A}=2\sin n\text{ A}\cos\text{ A}=\sin n\text{ A}\times\frac{\sin 2\text{ A}}{\sin\text{ A}}$$

from which it may be easily deduced.

The sines of the Surya Siddhanta, computed by the rules referred to, are given in the accompanying table, and they are compared with corresponding sines computed from modern tables.

The versed sines are computed by simply subtracting the sines of the complementary arcs from the radius 3438.

Table of natural sines and versed sines from Surya Siddhanta:—

Arc.	Indian Natural Sine R=3438.	Natural Sine from English Tables R=3438.	Versed Sines R=3438.
3° 45′	225	224·85	7
7 30	449	448·95	29
11 15	671	670·72	66
15 0	890	889·82	117
18 45	1105	1105·01	182
22 30	1315	1315·05	261
26 15	1520	1520·58	254
30 0	1719	1719·00	460
33 45	1910	1910·05	579
37 30	2093	2092·09	710
41 15	2267	2266·08	853
45 0	2431	2431·01	1007
48 45	2585	2584·08	1171
52 30	2728	2727·55	1345
56 15	2859	2858·55	1528
60 0	2978	2977·04	1719
63 45	3084	8083·45	1918
67 30	3177	3176·06	2123
71 15	3256	3255·75	2333
75 0	3321	3320·85	2548
78 45	3372	3371·95	2767
82 30	3409	3408·75	2989
86 15	3431	3430·85	3213
90 0	3438	3438·00	3438

F 2

The simple relations between the sine, cosine, and versed sine, served all the purposes of the trigonometry of the Hindu mathematicians, and though by name, as a function, distinct from signs, the tangent did not enter into their calculations; yet the ratio of the sign of an arc to the sine of its complement, or $\frac{\text{Sin A}}{\text{Sin }(90-\text{A})}$ the equivalent of tangent A, was of constant occurrence in their computations. For example, in the relation between the Gnomon and its shadow, the ratio, $\frac{\text{Gnomon}}{\text{Shadow}}=\frac{\text{Sin altitude}}{\text{Sin}(90\text{-alt.})}=\text{tan alt.}$

The Hindu astronomers appear to have been familiar with most of the elementary trigonometrical relations, such as

$\text{Cos A}=\sqrt{R^2-\sin^2 A}$, versed sine $A=R-\cos A$, $\sin 30°=\frac{R}{2}$

$\text{Sin } 45°=\frac{R}{\sqrt{2}}$, $\sin 18°=\frac{\sqrt{5R^2}-R}{4}$, $\sin 36°=\sqrt{\frac{5R^2-\sqrt{5R^4}}{8}}$

all of which are expressed in words, as also the more general relations between the sine and versed sine of an arc and the sine of half the arc, such as

$\text{Sine } \frac{A}{2}=\frac{1}{2}\sqrt{\sin^2 A+\text{versed }^2\text{sine } A}$, and $\sin \frac{A}{2}=\sqrt{\frac{1}{2} R \text{ versed sine } A}$

and hence as observed by Bhaskara.

"From the sine of any arc thus found, the sine of half the arc may be found (and so on with the half of this last). In like manner from the complement of any arc may be ascertained the sine of half the complement (and from that, again, the sine of half the last arc).

"Thus (Bhaskara says) the former astronomers prescribed a mode for determining the other sines (from a given one), but I now proceed to give a mode different from that stated by them."

And he gives, in words, rules which are equivalent to the well-known forms:

$\text{Sin}\left(45+\frac{A}{2}\right)=\sqrt{\frac{R^2+R\sin A}{2}}$ and $\sin\left(45°-\frac{A}{2}\right)=\sqrt{\frac{R^2-R\sin A}{2}}$

in which A is any arc of a circle.

And, again, when A and B are two arcs:

$$\text{Sin}\ \frac{A-B}{2} = \frac{1}{2}\left\{(\sin A - \sin B)^2 + (\cos A - \cos B)^2\right\}^{\frac{1}{2}}$$

"I will now give (he says) rules for constructing sines without having recourse to the extraction of roots." And the first of these is equivalent to

$$\text{Sin}\ (2\ A - 90^\circ) = \frac{R^2 - 2\sin^2 A}{R}.$$

"In this way several sines may be found." But this method, called Pratibhagajyaka-Vidhi, is limited. He then proceeds to give rules for finding the sine of every degree from 1° to 90°.

The rule for finding the sines of 36°, and of 18°, as given in Bhaskara's own words, are:—

"Deduct the square root of five times the fourth power of radius from five times the square of the radius, and divide the remainder by eight; the square root of the quotient will be the sine of 36°.

"Deduct the radius from the square root of the product of the square of radius and five, and divide the remainder by four; the quotient thus found will give the exact sine of 18°."

The first application in the Siddhanta showing the use of the sines is given in the form of a problem, thus:

Having given the longitude of a planet, to find its mean declination.

The mean declination of a planet is the same as that of the sun. when both have the same longitude, and the greatest mean declination is the same as the sun's greatest declination.

The rule given for the solution of the problem is:—Multiply the sine of the longitude by 1397, the sine of the (mean) greatest declination, and divide the product by the radius, 3438; find the arc whose sine is equal to the quotient; this arc is the planet's (mean) declination.

The rule given is evidently the same as that for finding the declination of the sun, when the longitude is given.

Thus, in the right-angled spherical triangle S ♈ N, in which S is

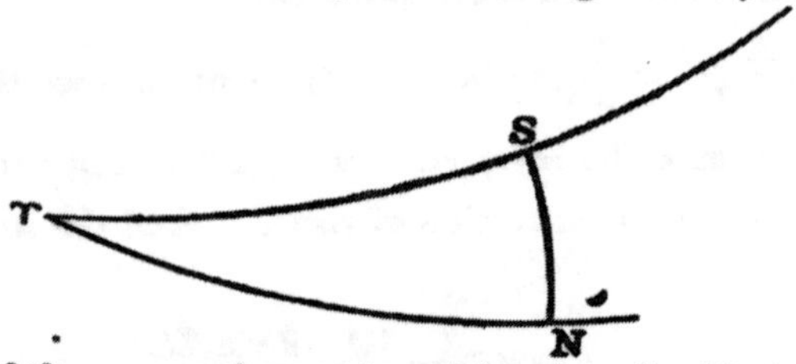

the place of the sun, ♈ the equinoctial point, S ♈ N the obliquity, and S N the sun's declination, we have

♈ sin S N = sin S ♈ N .Sin ♈ S.

But the sine S ♈ N = sin sun's greatest declination = 1397.

∴ 3,438 sin sun's declination = 1397 sin longitude.

Here it may be observed that 1,397 is also given in other places of the Surya Siddhanta as the sine of the sun's greatest declination.

It is said to be the sine of 24°, but by an accurate computation 1397 is the sine of the arc 23° 58′ 31″.

ON THE RULES FOR FINDING THE TRUE PLACE OF A PLANET.

The mean place of a planet, and that of its apogee, having been computed by the rules of the first chapter, the difference between them (called the Kendra, or mean anomaly) is taken, and the sine of it is found from the tables; these are used for constructing the various epicycles.

To explain the method of construction, let it be supposed that the circle of A B C D represents the deferent of a planet, in the

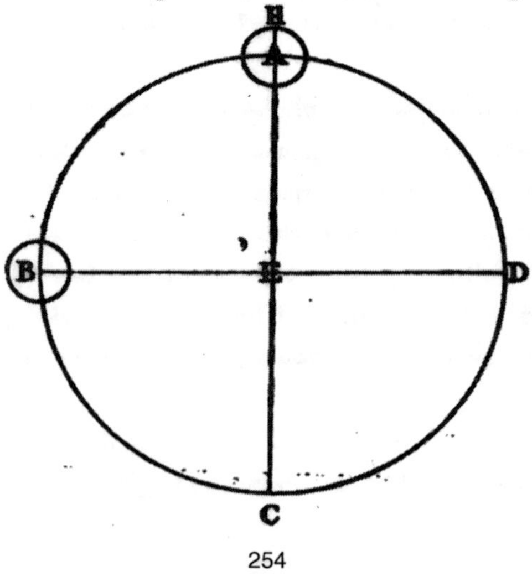

plane of the orbit, the radius being 3438, the centre E representing the centre of the earth, and the line A E C the line of Apsides.

Also let the small circle at A cutting the line of Apsides in H represent the epicycle of a planet, the point H of the circumference being at the greatest distance from E will be the apogee or higher apsis.

If we now conceive that whilst the centre of the epicycle (starting from A and supposed to be moving in the circumference of the deferent in the direction A B C D, *i.e.*, in the order of the signs of the Zodiac) completes one revolution in the deferent; the planet starting from H and moving in the epicycle also completes one revolution in the same time, in a contrary direction to that of the signs.

Then, it is assumed as a first approximation, that the direction in which the planet would appear to be seen at any time, when viewed from E, would be the direction of its true place.

The conception here formed is that of the ordinary epicycle, of an invariable magnitude. But the Indian epicycle has its circumference continually varying, being greatest when the centre is at A or C, and least when the centre is at B or D.

For the formation of the first epicycles of the sun and moon the following rule is given:—

(34) "There are fourteen degrees (of the concentric) in the periphery of the Manda, or first epicycle, of the sun, and thirty-two degrees (in the periphery of the first epicycle) of the moon, when these epicycles are described at the end of an even quadrant (of the concentric or on the line of the Apsides). But when they are described at the end of an odd quadrant (of the concentric or on the diameter of the concentric perpendicular to the line of Apsides), the degrees in both are diminished by twenty minutes."

The meaning of this rule is that the length of the circumference of an epicycle at A or C if stretched along A B C of the deferent would extend over an arc of 14° or 840′ for the epicycle of the sun, and for the moon the length would extend over 32° or 1,920′. But at the points B and D the lengths would be 820′ and 1,900′ respectively.

For an intermediate point M, the Kendra (or mean anomaly) of which was k, the peripheries of the epicycle of the sun would be

$$840' - 20 \times \frac{\sin k}{3438}, \text{ and that of the moon } 1{,}920' - 20 \times \frac{\sin k}{3438}.$$

In general, supposing C_A and C_B to be the circumferences of the epicycle, for any of the planets at the points A and B, respectively, then, for any intermediate point, M, with anomaly k the circumference of the epicycle would be

$$= C_A - (C_A - C_B) \frac{\sin k}{3438} = C_M,$$

such a circumference is called the (Sphuta or) rectified periphery.

After the method of constructing the rectified periphery of a planet's epicycle has been described, the following rule is given, for calculating the first equation of the planet, when the Kendra, or mean anomaly, is known:—

(39) "Multiply the sines of the Bhuja and Koti (of the first and second Kendra of a planet) by the rectified periphery (of the first and second epicycle of the planet); divide the products by the degrees in a circle (or 360°) (the quotients are called the first and second Bhuja-Phala and Koti-Phala, respectively). Find the arc whose sine is equal to the first Bhuja-Phala; the number of minutes contained in this arc is the Manda-Phala (or the first equation of the planet)."

To give an interpretation of this rule and the terms employed in it. it will be necessary to have recourse to a figure.

Let H, the intersection of the epicycle, with the line of Apsides E A, be the higher Apsis or Apogee. Then a planet is supposed to start from H, in the epicycle, with a uniform motion, and to describe the circumference of the epicycle, in the same time as its

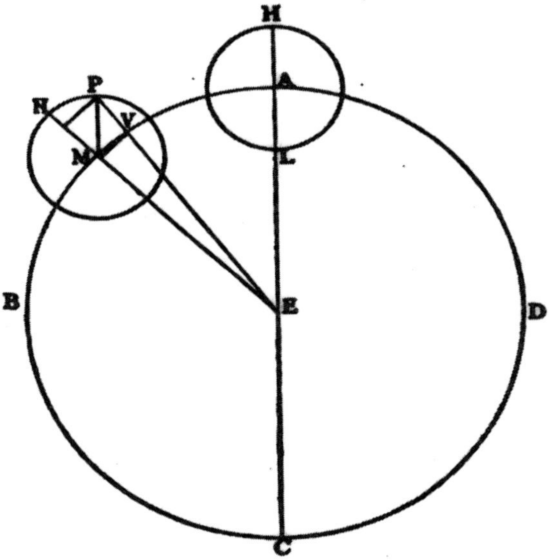

centre, starting from A, and moving uniformly in the concentric, completes its revolution in that circle.

When the centre has moved over an arc A M of the concentric, in the direction of the signs of the Zodiac, the planet P will have moved over an arc H P of the epicycle, such that, A M and H P will be similar arcs; therefore, the angle M E A will be equal to the angle H M P, so that P M will always be parallel to A E.

If, now, P E be joined, cutting the concentric in the point V, then the direction in which the planet would be seen, when viewed from E, the centre of the earth, would be that of the line E V, and the apparent place of the planet in the concentric would be V. Consequently, V is called the true place, and M V, the distance between the mean and the true places, is the arc required in finding the true place, and the object of the several rules for finding the equations of the centre.

As a first approximation, the passage quoted from the Surya Siddhanta (Rule 39) takes P n the perpendicular from P upon the line E M, as the sine of the Manda-Phala or first equation; this line is greater than the sine of M V, when the anomaly is less than 90°, and it is less than sin M V, when the anomaly is between 90° and 270°.

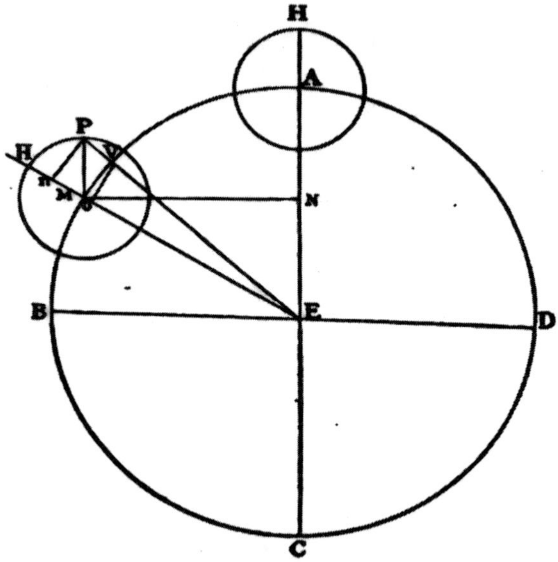

In the similar triangles P n M and M N E,

$$\frac{P\,n}{P\,M} = \frac{M\,N}{M\,E}, \text{ or } P\,n = \frac{P\,M}{M\,E} \cdot M\,N,$$

but the circumferences are as the radii,

$$\therefore \frac{P\,M}{M\,E} = \frac{\text{circumference of epicycle at M}}{\text{circumference of concentric}},$$

and M N is the sine of the arc M A = sin k.

$$\therefore P\,n = \frac{C_M}{21{,}600'} \cdot \sin k.$$

Also from the same similar triangles,

$$\frac{M\,n}{M\,P} = \frac{N\,E}{M\,E}, \text{ or } M\,n = \frac{M\,P}{M\,E} \cdot N\,E,$$

in which N E is the cosine of the arc A M.

$$\therefore M\,n = \frac{C_M}{21{,}600'} \times \cos k.$$

If now, in the figure, we call M N the Bhuja, and P n the Bhuja-Phala, also N E the Koti, and M n the Koti-Phala, and we express the ratio $\frac{C_M}{21,600}$ in degrees instead of minutes, we shall have in the words of the rule,

$$\text{Bhuja-Phala} = \frac{\text{rectified periphery at M}}{360^\circ} \times \text{Bhuja, and}$$

$$\text{Koti-Phala} = \frac{\text{rectified periphery at M}}{360} \times \text{Koti.}$$

And the arc of which Bhuja-Phala is the sine is the Manda-Phala, or first equation of the centre.

This last would be correct, if P E had been the radius. It is partly the object of Rules 40, 41, 42 to make this correction.

FOR THE GREATEST EQUATIONS OF THE SUN AND MOON.

If we substitute for C_M in the equation $P\,n = \frac{C_M}{21,600} \times \sin k$, the rectified peripheries of the sun and moon, respectively, we have for the sun,

$$\text{the sine first equation of the sun} = \frac{840 - 20 \times \frac{\sin k}{3438}}{21,600} \times \sin k.$$

This will have its greatest value when $k = 90^\circ$, and $\sin k = 3438$,

$$\therefore \text{Sin greatest first equation of the sun} = \frac{820}{21,600} \times 3438 = \sin 2^\circ\, 10'\, 32''$$

Or the greatest equation of the centre $= 2^\circ\, 10'\, 32''$.

At the beginning of the present century, according to Laplace, the greatest equation of the sun's centre was 1° 55′ 27·7″, and this diminished at the rate of 16·9″ in a century; the difference between these two values of the greatest equation = 904·3″, which, divided by 16·9″, gives 5,351 years as having elapsed up to the beginning of A.D. 1800, since the greatest equation of the centre had the value given to it in the Surya Siddhanta.

When the moon's rectified periphery C_M is substituted the sine of the moon's first equation $= \dfrac{1,920 - 20 \dfrac{\sin k}{3,438}}{21,600} \times \sin k$, which, when $k = 90°$ and $\therefore \sin k = 3,438$, becomes the sine of the moon's greatest equation $= \dfrac{1,900}{2,160} \times 3,438 = \sin 5° \; 2' \; 47''$, or the moon's greatest equation $= 5° \; 2' \; 47''$.

At the beginning of this century, according to Laplace, the moon's greatest equation was 6° 17′ 54·5″. But he adds: "The constant effect of the evection is to diminish the equation of the centre in the syzigies, and to augment it in the quadratures; at its maximum it amounts to 1° 18′ 2·4″."

Thus, since the Hindus were not acquainted with the evection, as a term distinct from the equation of the centre, the value which they give to the moon's equation was not inconsistent with that given by Laplace.

The rules already referred to, as giving a more correct value of the equation, are put in the following order:—

(40) "To find the second equation of the minor planets—Mars, etc. Find the second Koti-Phala (from a planet's second Kendra). It is to be added to the radius, when the Kendra is less than three signs or greater than nine signs; but when the Kendra is greater than three signs and less than nine, then the second Koti-Phala is to be subtracted from the radius.

(41) "Add the square of the result (just found) to that of the sine of the second Bhuja-Phala; the square of the sum is the Sighra-Karna, or second hypothenuse.

"Find the second Bhuja-Phala of the planet (as mentioned in Sloka 39); multiply it by the radius, and divide the product by the second hypotenuse (found above).

(42) "Find the arc whose sine is equal to the quotient (just found); the number of minutes contained in the arc is called the Sighra-Phala (or second equation of the planet)."

These rules may be explained in the following manner:—

The figure being the same as that given before, in the first rule (39) we have, as in that case

$$P\,n = \frac{C_M}{21600} \cdot \text{Sin } k = \text{Bhuja-Phala.}$$

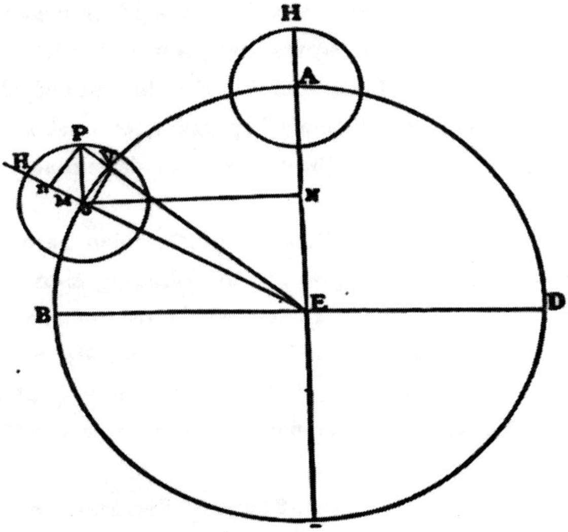

In which *k* may be the first or second Kendra, and according to the position of M, sin *k* may make M *n* vary in sign.

$\therefore$ E *n* may be taken $=$ E M $\pm$ M *n*, but in the right-angled triangle P *n* E, P E^2 $=$ P *n*^{2} $+$ E *n*^{2}, or

$$\text{Sighra-Karna} = P\,E = \sqrt{\{P\,n^2 + (E\,M \pm M\,n)^2\}}$$

$$= \sqrt{\left\{\left(\frac{C_M \sin k}{21600}\right)^2 + \left(3438 \pm \frac{C_M \cos k}{21600}\right)^2\right\}}$$

Again, in the similar right-angled triangles P *n* E and V *o* E

$$\frac{V\,o}{P\,n} = \frac{V\,E}{P\,E}, \text{ or } V\,o = \frac{V\,E \cdot P\,n}{P\,E}$$

$$\therefore \text{Sin } O\,V = \frac{3438 \times \dfrac{C_M \sin k}{21600}}{\sqrt{\left\{\left(\dfrac{C_M \sin k}{21600}\right)^2 + \left(3438 \pm \dfrac{C_M \cos k}{21600}\right)^2\right\}}}$$

If we express this formula in words with Indian names, we have the rule (40–42)

$$\text{Sin Sighra-Phala} = \frac{\text{Radius} \times \text{Bhuja-Phala}}{\text{Karna}}.$$

$$\text{Where Karna} = \sqrt{\{(\text{Radius} \pm \text{Koti-Phala})^2 + \text{Bhuja-Phala}^2\}}.$$

There is no difficulty in forming tables from either of the two rules contained in (39–42), when the Kendra of M, or mean anomaly, is taken for each of the 24 sines, that are given in the table of sines, or even for every degree of a quadrant. For the sun and the moon, which have each only one system of epicycles, one set of tables is deemed sufficient for each of these bodies, to determine their true places.

But this is not the case with the planets, which have each two distinct systems of epicycles, given in the following form :—

(35) "There are 75, 30, 33, 12 and 49 degrees of the concentric in the peripheries of the first epicycles of Mars, Mercury, Jupiter, Venus and Saturn, respectively, at the end of an even quadrant of the concentric, but at the end of an odd quadrant there are 72, 28, 32, 11, 48 degrees of the concentric.

(36) "There are 235, 133, 70, 262 and 39 degrees of the concentric in the peripheries of the Sighra, or second epicycles, of Mars, etc., at the end of an even quadrant (of the concentric).

(37) "At the end of an odd quadrant (of the concentric) there are 232, 132, 72, 260, 40 degrees of the concentric in the peripheries of the second epicycles of Mars, etc."

The rectified peripheries of the planets are formed exactly in the same way as that of the sun.

Taking Mars as an example, its first epicycle, or Manda, would be, for a point M, whose Kendra was k, using the same notation as before,

$$C_M = C_1 - (C_1 - C_2)\frac{\sin k}{3438},$$

which, expressed in minutes,

$$= 4500 - 180 \cdot \frac{\sin k}{3438}.$$

And for the second epicycle, or Sighra, the rectified periphery would be for a Kendra *k*,

$$C_M = 14100 - 180 \cdot \frac{\sin k}{3438}.$$

If, now, these values be put successively in the rules (39-42) for the rectified periphery, they will afford the means of calculating two tables for Mars; the one being a table for the first equation of the centre, or Manda-Phala, and the other table will give the Sighra-Phala, corresponding to the Indian estimate of the aunual parallax of the planet.

When these tables have been formed there is still one more rule by which they are to be applied, so as to find the true place of a planet. It is the following:—

(44) "Find the equation (from the mean place of a planet) apply the half of it to the mean place, and (to the result) apply the half of the first equation (found from that result), from the amount find the first equation again, and apply the whole of it to the mean place of the planet, and (to that rectified mean place) apply the whole of the second equation found from the rectified mean place; thus you will find the true place of the planet."

This method of approximating to the true place of a planet, by successive steps, would appear to resemble our method of approximating to the length of a small curve, by supposing it to lie between the lengths of its chord and tangent.

When the mean place is in advance of the true place the equation is to be subtracted, but when the true place is in advance of the mean place the equation is to be added, and at the higher and lower apsides the two places will be coincident.

The rule for the first equation in the Surya Siddhanta would appear to have formed the subject of a discussion among the astronomers, with reference to the reason why the Karna, or hypotenuse, should have been omitted.

Bhaskara says: "Some say that the hypotenuse is not used in the first process, because the difference is inconsiderable, but others maintain that since in this process the periphery of the first circle, being multiplied by the hypotenuse and divided by the radius becomes true, and that, if the hypotenuse then be used, the result is the same as it was before, therefore the hypotenuse is not employed. No objection is made why this is not the case in the second process, because the proofs of finding the equation are different here."

He gives only a very brief account of the method of the Epicycle, and he seems to make the circle invariable in magnitude, which would be consistent with the method of the eccentric, to which he appears to have given the preference.

ON THE METHOD OF THE ECCENTRIC.

Let A B C D and H P L be two circles having their radii each equal to 3438′, the number of minutes in the radius of the table of

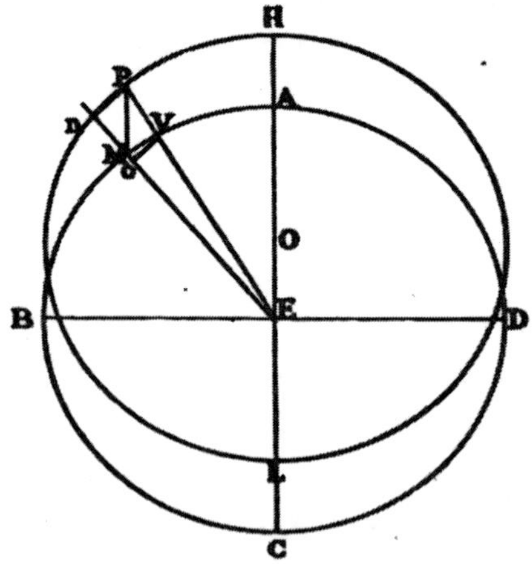

sines, and their centres E and O at a distance from each other equal to the number of minutes in the greatest equation of the centre of a planet's orbit.

If E represent the centre of the earth, then the circle A B C D is called the concentric and H P L whose centre is O is called the

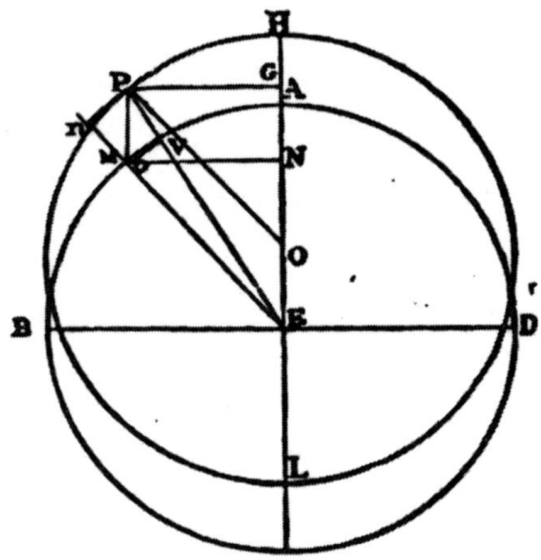

eccentric. The line joining O E and produced, is the line of apsides. The point H, where it meets the eccentric is the higher apsis, and L the lower.

Let it now be conceived that the planet, moving in the eccentric describes the arc H P, in the same time that an imaginary planet moving in the concentric with the same mean motion, describes the arc A M, then the arcs A M and H P being equal, the radii, M E and P O are parallel, and the lines M P and E O joining equal and parallel lines are themselves equal and parallel, therefore P M the line joining the true and imaginary planet is always equal and parallel to the eccentricity E O.

Now the line E P which joins the centre of the earth with the planet, is the direction in which the planet will be seen from E, and V the point in which it meets the concentric is called the true place of the planet: the distance M V between the mean and true places is the equation of the centre.

Q

The arc M A is the Kendra or mean anomaly $= k$, and M N drawn from M perpendicular to H E $=$ sin M A $=$ sin k, N E is the cosine of M A $=$ cos k.

In the similar triangles M P n and P E G, $\frac{Mn}{MP} = \frac{PG}{PE}$ or

$$Mn = \frac{MP \cdot PG}{PE} = \frac{e \sin k}{PE}, \; e \text{ being the eccentricity}$$

$$\text{Also } PE = \sqrt{PG^2 + EG^2} = \sqrt{\{\sin^2 k + (\cos k \pm \epsilon)^2\}}$$

$$\therefore \text{Sin equation of the centre} = Mn = \frac{\epsilon \sin k}{\sqrt{\{\sin^2 k + (\cos k \pm \epsilon)^2}}$$

The terms used in the rules to designate the same lines, are not always the same, thus in the formula for the equation of the centre just given: Sin k is the line M N or its equal P G of the figure. It is sometimes called "the sine of the Bhuja of the Kendra," at other times it is called "the Bhuja of the Kendra," and again more briefly "the Bhuja." Cos k which is the line N E or its equal G O is sometimes called "the Koti" at other times "the sine of the Koti," and more at length "the sine of the Bhuja of the complement of the Kendra." Again, cos $k + \epsilon$ is the line E G, or E N + N G. it is called the Sphuta Koti, the line P E is the Karna or hypotenuse, and we have Karna^2 = Bhuja^2 + Sphuta Koti^2.

The term sine Bhuja which is applied here so frequently, belongs, properly, to an arc of the concentric, or deferent. This term will be easily understood as being the numerical valuer of such quantities as sin (90 + A), sin (180 ± A), etc., the Bhuja being the equivalent arc for the table of sines.

The sines and cosines designated with such names, are lines in the circle whose radius is 3438. When similar sines and cosines are taken in the circle whose radius is ϵ, which is that of the Epicycle, variable or invariable, the sine becomes Bhuja Phala and the cosine is Koti Phala.

That the two methods of calculating the equation of the centre, whether by the epicycle or the eccentric, lead to the same result, is thus indicated by Bhaskara.

"If the diagrams (of the eccentric and epicycle) be drawn unitedly, and the place of the planet be marked off, in the manner before explained, then the planet will necessarily be in the point of intersection of the eccentric by the epicycle."

If we draw first the figure for the eccentric and concentric, using the same letters for the several lines as before, then the planet moving in the eccentric, and starting from H the higher apsis, is supposed to have described the arc H P in the direction of the signs, and the line P M drawn parallel to H E, will meet the concentric in the point M, the place of the imaginary mean planet supposed to be moving in that circle, and P M has been shewn to be equal to E O, the eccentricity.

If now at M, as a centre, with a radius equal to the eccentricity, a circle be described, this circle will be the epicycle, in which a planet moves through the arc H P from the higher apsis H, in the same time that its centre will have described the arc A M of the concentric, and the radius being equal to the eccentricity it is equal to H P, or the epicycle and the eccentric intersect in the same point, and all the other lines and angles of the figures, according to one method are identical with those of the other method.

It is also obvious that in the eccentric the planet will move on the arc H P of the eccentric, in the direction of the signs of the Zodiac; but in the epicycle, the arc H P of that circle is in a direction opposite to the signs.

Eccentricities of the orbits of some of the planets given in the Sidhantas as compared with corresponding values of the eccentricities, in modern tables, are as under:—

	Eccentricity in parts of radius of deferent.	Eccentricity in terms of the semi-axis.—*Herschel's Astronomy.*
The Earth ...	·01944 and ·019	·016783
The Moon ...	...	
Mercury... ...	...	·2055149
Venus	...	·00686
Mars	·10416 and ·1	·0933
Jupiter	·04583 and ·0444	·04816
Saturn	.068 and ·0666	·05615

Q 2

Rule (45), has reference to the application of the 1st and 2nd equations—whether they are additive or subtractive, and (46) relates to a correction to the places named Bhujantara.

Rules from (47) to (51), give methods of finding the true diurnal motions of the Sun, Moon and Planets.

Rule (52), explains the Indian theory of the retrogression of the planets, supposed to be caused by loose reins of attraction.

Rules (53) to (55), detail the conditions on which the motion of the planets Mars, Mercury, Jupiter, Venus, and Saturn begin to be retrograde, as occurring when their Kendras or Anomalies are respectively 164°, 144°, 130°, 163° and 115°, and that the retrograde motion ceases at 196°, 216°, 230°, 197° and 245°.

Rules (56) and (57), give an Indian method of calculating the latitude of a planet, by a species of proportion, supposing the difference between the rectified places of the planet, and its node and the greatest latitude of the planet to be all known.

Rule (58), makes a distinction in computing the declination of the Sun, and that of a planet, the former being the true declination of the place of the Sun in the ecliptic, and the latter, called the mean declination of the planet, is computed from the planet's place, referred to the ecliptic, increased or diminished by its *latitude* north or south.

Rule (59), derives the length of a planet's day and night, by proportion, from its diurnal motion, reduced to an arc of right ascension in time.

Rule (60), makes the radius of the diurnal circle of a planet=3438 – versed sine declination.

Rule (61), determines the ascensional difference of a planet, by a somewhat operose method, by means of the radius of the planet's diurnal circle, its declination, and the equinoctial shadow (described in the third chapter).

Rule (62)—(63), give again a loose method of finding the lengths of a planet's day and night, by a reference to the ascensional difference found by the preceding rule.

From Rule (64) is found the Nacshatra in which a planet is at a given time, together with the computation of the days and parts of a day from its entrance into that Nacshatra.

Rule (65), to find the Yoga (an astrological period in which the sum of the places of the Sun and Moon, increases by 13°″20′) at a given time, together with the number of elapsed Yogas (counting from the one named Vishkarubha).

Rule (66), gives a method of finding the lunar day at a given time.

The remainder of the chapter from (67) to (69) relates to certain portions of time, called the 4 invariable and the 7 variable Karanas which are said to answer successively to the half of a lunar day.

CHAPTER III.

RULES FOR RESOLVING QUESTIONS OF TIME, ETC.

The Hindu astronomer has various simple means of observation, more or less effective in regard to accuracy. His observatory, if we might give it so dignified a name, was homely, but adapted for furnishing him with the data on which his calculations were made.

It consisted principally of a levelled horizontal plane, a floor or terrace of chunam, which is a lime made from shells, and which, when dry, is hard and capable of receiving a polish equal to that of marble.

At a point of the floor as a centre, a circle is described, and a fine vertical rod of given length is erected at this point, as a stile or Gnomon, and by means of the length and direction of its shadow cast on the plane by the Sun, a variety of astronomical problems are solved.

Such problems appear in Chapter III. of the Surya Siddhanta treating of questions relating to time, position of the heavenly bodies and their directions.

Rules (1-4) give the method of drawing a meridian line, and the east and west line on a horizontal plane.

On the surface of the stone or chunam floor, levelled with water, a circle is described, and at the centre a vertical Gnomon is placed whose length is (12) digits: in the morning and afternoon, the two points, where the shadows of the Gnomon meet the circumference, are marked, from these two points as centres, intersecting arcs are described; the point of intersection is called the Timi (the fish, named from its form). A line is then drawn from the Timi through the centre of the circle, this is called the north and south line, or meridian line, and the line through the centre at right angles to the N S line is the east and west line.

Our works on dialling and astronomy give the same method of drawing a meridian line on a horizontal plane.

Rule (5) directs a circle to be described on the horizontal plane, with a radius equal to the shadow of the Gnomon, and a square to be described about the circle, the sides touching it at the four cardinal points.

If the figure represent the circle, with the circumscribed square, and O H the shadow of the Gnomon, a perpendicular H B is then drawn to the E W line, in this case before noon, and the sun having

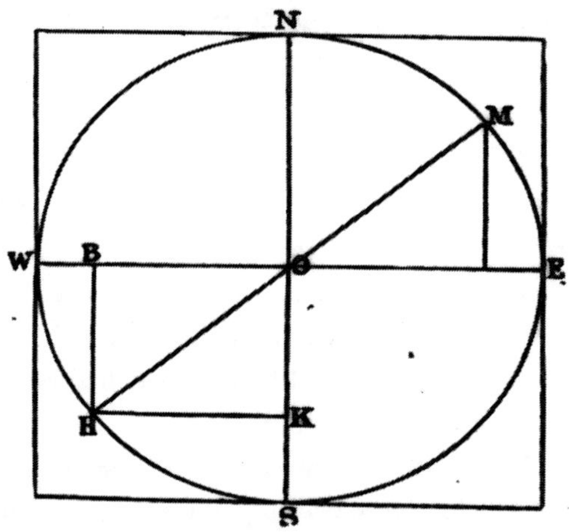

a north declination. Then H B is called the Bhuja (or sine) of the shadow and H K, or its equal B O, is called the Koti.

According to the position of the end of the shadow, the Bhuja is distinguished as being north or south, and the Koti as being east or west.

The direction of the sun being in the vertical plane of the Gnomon and its shadow, the shadow produced backwards, will be the line of intersection of this vertical plane with the plane of the horizon, and the angle E O M, or its measure the arc E M, will be the amplitude of the sun at the given time, and M N will be the measure of its azimuth.

Rule (6) merely states that the three circles of the sphere, the prime vertical, the equinoctial, and the six o'clock circle pass through the east and west points of the horizon.

(7) This rule directs, that in the circle before described a line is to be drawn parallel to the E W line, at a distance from it equal to the length of the equinoctial shadow; it is then stated, that the distance between the end of the given shadow and the latter line, is equal to the sine of amplitude (reduced to the hypotenuse of the given shadow).

In this rule two important points need explanation, first, with regard to the equinoctial shadow, and the line drawn at a distance equal to its length on the plane, parallel to the east and west line.

The definition of the equinoctial shadow is given afterwards in rule (12), as the shadow cast upon the meridian line of a given place at noon, when the sun is in an equinox.

It is called the Palabha, and it is a primary constant, in problems which involve the latitude of the place; for a ray of light from the equinoctial Sun at noon makes, with a vertical line of any place, an angle equal to the latitude. Hence we have—

The length of the equinoctial shadow = Gnomon × tan latitude.

But since the Hindus made use only of sines and cosines,

$$\text{The Palabha} = \text{Gnomon} \times \frac{\text{sin latitude}}{\text{cos latitude}}$$

The properties of the line drawn parallel to the E W line in rule (7), may be thus explained.

Let N P Z S represent the meridian of any place, N E S W the horizon, E Q W a portion of the celestial equator, and Z the zenith of the place.

Conceive a plane *e* q *w* parallel to the plane of the equator to pass through the end G of a vertical Gnomon O G, and to meet the horizon, the intersection of the two planes will be a straight line *e* A *w*. Suppose now when the sun is in either equinox, all

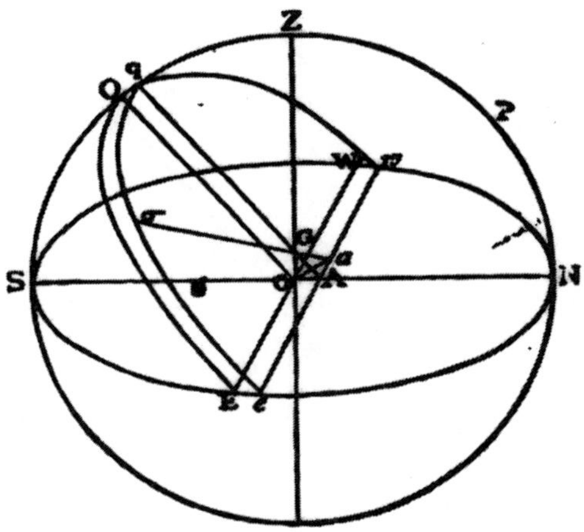

rays from it, in its daily course passing through the point G, will lie in the surface of the plane *e* q *w*, and if *σ* G *a* be any one such ray it will meet the horizontal plane in the line of intersection of the two planes *e* A *w*, and for this position of the sun O *a* will be the shadow cast by O G, and, throughout the day of the equinox, the ends of all the shadows will be points of the line *e* A *w*, but O A, the shadow when the equinoctial sun is in the meridian is the one which obtains the name of the equinoctial shadow.

Secondly. The other point of rule (7) needing explanation, is the proposition that the distance between the end of a given shadow and the line of the equinoctial shadows, is equal to the sine of the amplitude (reduced to the hypotenuse of the given shadow).

In the following figure let the circle as before described on the horizontal plane be N E S W, and M O H R the intersection on the horizon of a vertical circle through the sun at any time, H O the shadow of the Gnomon, and *e w* the line drawn parallel to the E W line, at a distance equal to O A, the equinoctial shadow; then, if from H the end of the given shadow, the line H L T be drawn perpendicular to the line *e w*, the proposition states that H T is equal to the sine of the amplitude reduced to the hypotenuse. Rule (7) only enunciates

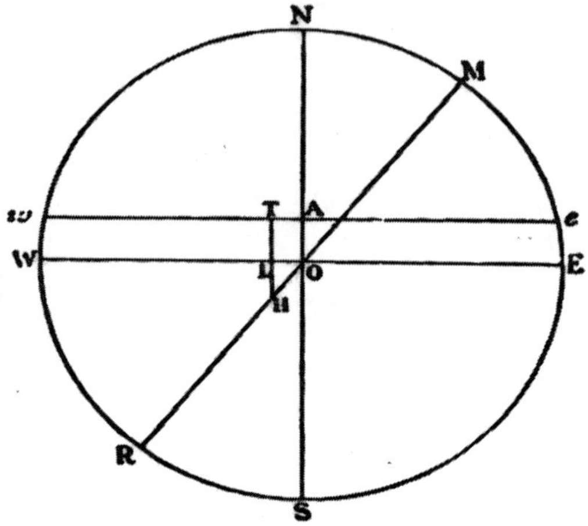

this singular property, the proof is given afterwards by successive steps in rules (21-24).

Rule (8) deduces the hypotenuse of the right-angled triangle, of which the Gnomon of 12 digits and the shadow form the sides.

The shadow being given, the hypotenuse $= \sqrt{(144+S^2)}$.

Rule (9) has been referred to before as stating "that the circle of Asterisms librates 600 times in a Great Yuga. The libration being performed through 27° from a mean point, in a retrograde direction; then returning to the mean point, proceeding through 54° with a direct motion, thence returning to the mean point with a retrograde motion: thus making 108° for a complete libration.

This hypothesis gives the mean annual precession of the equinoxes = 54", a quantity which is adopted by most of the other Siddhantas.

The rule further directs the calculation to be carried backwards proportionally, the quantity of it to be computed by means of the days in a Kalpa or a Yuga, and the days which have elapsed from the assumed epoch.

Rule (10). The amount of the precession is to be applied to the place of a planet and from the result the declination, the shadow of the Gnomon, the ascensional difference, etc., are to be computed.

Rule (11) has reference to the computed true place of the sun, found in the second chapter, as compared with the place found by observation.

Rule (12) has been before referred to (Rule 7) as giving the definition of the equinoctial shadow, called the Palabha.

In rule (13) the converse problems are proposed; to find from the equinoctial shadow, the latitude and the co-latitude.

The hypotenuse of the equinoctial shadow being known, then

$$\text{Cos } l = \frac{12\text{ R}}{h}, \text{ and sin } l = \frac{\text{R}}{h} \times \text{Palabha.}$$

Where $\text{R} = 3438$, the number of minutes in the arc of a circle whose length is equal to the radius of the circle, as mentioned before (p. 209), it is the length of an arc of 57° 18′, and is used in modern analysis, and styled the analytical unit.

The reverse operation gives

$$l = \sin^{-1} \frac{\text{Palabha}}{h} \times \text{R} = \sin^{-1} \frac{\text{S} \times \text{R}}{\sqrt{s^2 + 12^2}}$$

$$= \cos^{-1} \cdot \frac{12\text{ R}}{h} = \cos^{-1} \cdot \frac{12 \cdot \text{R}}{\sqrt{s^2 + 12^2}}$$

The Indian trigonometrical functions are circular functions, radius being 3438.

Rules (14) and (15). Suppose the shadow at noon to be given at any other time than at the equinox and the sun's declination to be known; to find the latitude it is obvious that the vertical angle of the right-angled triangle of which the Gnomon and its shadow are the sides, is always the angular zenith distance of the direction of the hypotenuse, and $\sin z = \frac{\text{R} \cdot \text{S}}{\text{H}}$, where S is any shadow and H the hypotenuse.

$$\text{Then, } z = \sin^{-1} \frac{\text{R S}}{\text{H}}.$$

The rule, moreover, states, that the sum, or difference of the sun's zenith distance at noon, and the declination is the latitude of the place, or, $l = z \pm d$.

Rule (16) deduces the equinoctial shadow, or Palabha, from a given latitude, and the

$$\text{Palabha} = \frac{12 \sin l}{\sqrt{R^2 - \sin^2 l}}$$

By rules (17-18) the sun's declination and his longitude are to be found, when his meridian zenith distance and the latitude of the place are given.

From (15) $d = z - l$ or $l - z$.

Let S be the place of the sun on the ecliptic ♈ ♋ ♎ at any time, and the arcs ♈ S, S B and B ♈ of the right angled spherical triangle ♈ S B

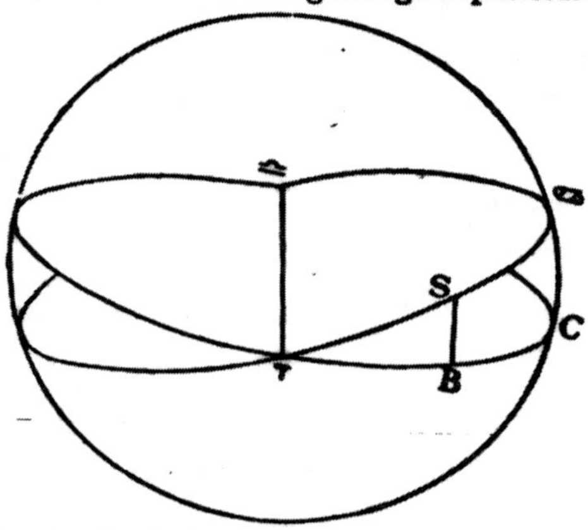

be the sun's longitude, declination and right ascension respectively. Then in this triangle we have R sin S B = sin ♈ S · sin S ♈ B.

$$\therefore \text{Sin longitude} = \frac{\text{R sin declination}}{\text{sin obliquity}}$$

But the ancient Hindu Astronomers made the obliquity 24°, the sine of which from their table is 1,397 to radius 3,438 minutes. Therefore, sin sun's longitude $= \frac{3438}{1397} \times$ sin sun's declination or

Sun's longitude $= \text{Sin}^{-1}\left(\frac{3438}{1397} \times \sin d\right)$, in the first quadrant of his orbit

$= 180° \mp \sin^{-1}\left(\frac{3438}{1397} \times \sin d\right)$ in the second and third quadrants, and

$= 360° - \sin^{-1}\left(\frac{3438}{1397} \times \sin d\right)$ in the fourth quadrant

Rule (19) corrects the sun's longitude to his place, as referred to the 1st point of Mesha or Aswini, by substracting the elapsed precession, from about 570 A.D., when the equinox was at the beginning of that sign or Nacshatra, calculated at a mean annual rate of 54."

The corrected place thus found is called the true place of the sun, from which by a process inverse to that described in the second chapter, the mean place is to be found. This is accomplished by a series of steps, first finding an approximate mean place by subtracting or adding the 1st equation, as in the table of equations from or to the true place. A nearer approximation is then found, by adding or subtracting the equation to or from the approximate mean place found before, and so on repeatedly till the exact mean place is found.

Rule (20). Supposes the latitude of a place and the sun's declination to be given, to find the zenith distance at noon, thus

$$z = l \pm d$$

From which $\text{Sin } z = \sin(l \pm d)$ and

$$\text{Cos } z = \sqrt{\left(R^2 - \sin^2(l \pm d)\right)}$$

Rule (21). Assuming the sun's zenith distance at noon to be given, to find the shadow of the Gnomon, and the hypotenuse of the triangle, having these lines as sides at noon

$$\text{Shadow at noon} = \frac{12 \sin(l \pm d)}{\sqrt{(R^2 - \sin^2(l \pm d)\}}} = 12\frac{\sin z}{\cos z}$$

$$\text{Hypotenuse at noon} = \frac{12\,R}{\sqrt{\{R^2 - \sin^2(l \pm d)\}}} = 12\frac{R}{\cos z}$$

Rule (22). The shadow and the sun's declination being given, to find his amplitude, and the sine of the amplitude reduced.

$\text{Sin amplitude} = \frac{h \sin d}{12}$, when h is the equinoctial hypotenuse.

The amplitude here is evidently that of the sun when rising.

Sin reduced amplitude = sin sun's amplitude $\frac{H}{R}$, H being the hypotenuse at noon on the given day

$$\therefore \left.\begin{matrix}\text{sin reduced}\\ \text{amplitude}\end{matrix}\right\} = \frac{h\,H}{12\,R}\sin d;\ \text{but at noon}\ \frac{H}{R} = \frac{12}{\cos(l \pm d}$$

$$= \frac{h \sin d}{\cos(l \pm d)}$$

Rule (23). Assuming the equinoctial shadow, and the sine of the reduced amplitude to be given, to find the Bhuja, the rule gives for this purpose the following cases: reduced sine of amplitude + Palabha = North Bhuja when d is south; reduced sin amplitude − Palabha = North Bhuja when d is north.

Rule (24) states reduced sin amplitude − Palabha = South Bhuja when d is south, and every day at noon the Bhuja is equal to the Gnomonic shadow at that time.

The truth of the property enunciated in rule (7), already referred to, may be proved by modern trigonometrical methods, the radius being assumed unity.

Taking the figure before given and described under rule (7), in which M O R represents the intersection of the horizon with the vertical circle passing through the sun at any time, and O H to be then the shadow of a Gnomon of 12 digits, cast on the plane:

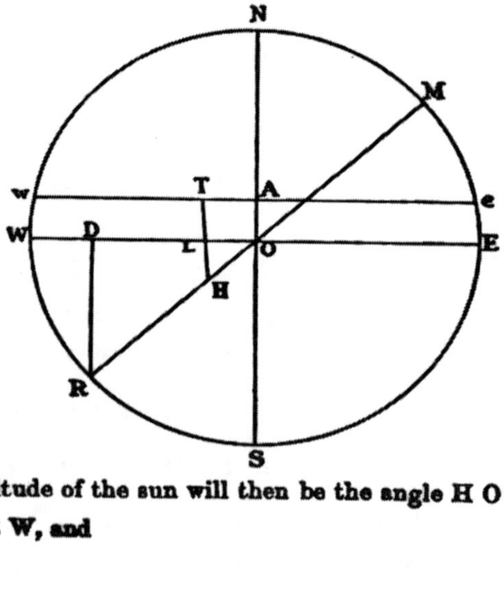

The amplitude of the sun will then be the angle H O L, measured by the arc R W, and

$$\frac{HL}{HO}=\frac{RD}{RO}=\sin ROD$$

$$\therefore HL = s \sin \text{sun's amplitude} \ldots \quad (1)$$

Again by spherical trigonometry it is easily shown that:

the sine of the sun's amplitude $= \dfrac{\sin d}{\cos l \sin z} - \tan l \cos z \ldots$ (2)

Where d = the sun's declination, z = his zenith distance and l = the latitude of the place.

But from the right angled triangle of which the Gnomon and shadow are represented by 12 digits and S, and the hypotenuse by H.

From rule (12) we have $\dfrac{\sin l}{\cos l} = \dfrac{\text{Palabha}}{12} = \tan l$

and from (20), $\sin z = \dfrac{S}{H}$ and $\dfrac{\cos z}{\sin z} = \dfrac{12}{S} = \cos z$

By substitution of these in equation (2) it becomes

$$\text{Sin sun's amplitude} = \frac{H}{S} \cdot \frac{\sin d}{\cos l} - \frac{\text{Palabha}}{S}$$

$$\text{or, } S \cdot \sin \text{amplitude} + \text{Palabha} = H \frac{\sin d}{\cos l} \ldots\ldots \quad (3)$$

Now, if A represent the amplitude, when the sun is just rising or setting on the same day

$$\sin A = \frac{\sin d}{\cos l}$$

By substitution in (3) we have

$$S \times \text{sine amp. at given time} + \text{Palabha} = H \sin A.$$

But in (1) $S \cdot \sin$ amp. = H L, and Palabha = L T, and their sum

$$HT = H \sin A.$$

Rule (25) assumes the latitude and sun's declination to be given; then, when the Sun is on the prime vertical, the hypotenuse of the shadow is found from

$$H = \frac{12 \sin l}{\sin d} \text{ or}$$

$$= \frac{\cos l}{\sin d} \times \text{equinoctial shadow}$$

The figure being supposed to represent a projection of circles of the sphere on the meridian, D *s* S U the diurnal path of the sun and S his place on the prime vertical O Z, and S P his distance from the

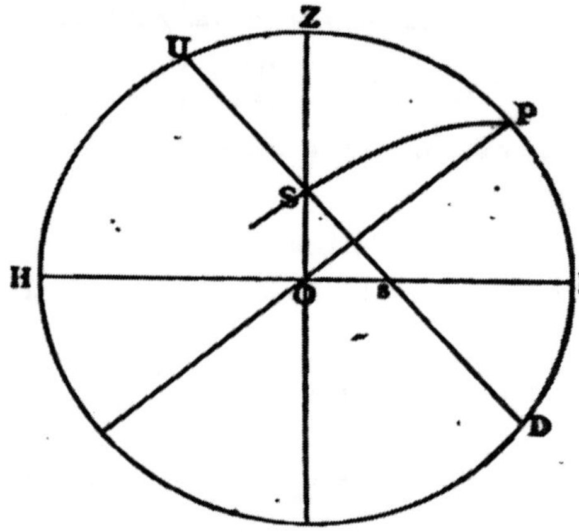

pole. Then in the spherical triangle P Z S, the angle P Z S is a right angle, and by Napier's rule R cos S P = cos S Z · cos Z P.

(1) or, $R \sin d = \cos Z \cdot \sin l$ where $S P = 90 - d$; $S Z = Z$ and $Z P = 90 - l$.

But in the triangle formed by the Gnomon, its shadow and the hypotenuse H, $\cos Z = 12 \frac{R}{H}$, substituting in (1) and reducing

(1) $H = \frac{12 \sin l}{\sin d}$, but $12 \sin l = \text{Palabha} \times \cos l$.

(2), also $H = \frac{\text{Palabha}}{\sin d} \cdot \cos l$

In rule (27), the value of H, is found otherwise, when d is north and less than l, from

(3) $H = \frac{\text{Palabha} \times h_1}{\text{reduced sin amplitude at noon}}$, h_1 being the hypotenuse at noon.

But in rule (22) it is shewn that the reduced amplitude at noon

$$= \frac{h_1}{R} \sin A, \text{ where A is the rising amplitude.}$$

And $\sin A = \frac{R \sin d}{\cos l}$, substituting these in (3)

$$H = \frac{\text{Palabha}}{\sin d} \times \cos l, \text{ the same as (2).}$$

Rules (28), (29) and (30) are preparatory to (31) and (32), the object of which is to find the sun's altitude when in the vertical circle whose azimuth is 45°.

In (28) and (29) the term Karani is assumed

$$= \frac{144\left(\frac{R^2}{2} - \sin^2 A\right)}{72 + \text{Palabha}^2}$$

In (30) the term Phala is assumed

$$= \frac{12\ P \cdot \sin A}{72 + \text{Palabha}^2}$$

In which Palabha is the equinoctial shadow and A is the rising amplitude of the sun.

In rules (31) and (32), the Kona-Sanku or sine of the sun's altitude when his azimuth is 45° $= \sqrt{\text{Karani} + \text{Phala}^2} \mp \text{Phala}$.

To prove the truth of this solution, in the figure, let the circle H Z P O represent the meridian; H R O the horizon; and Z S R

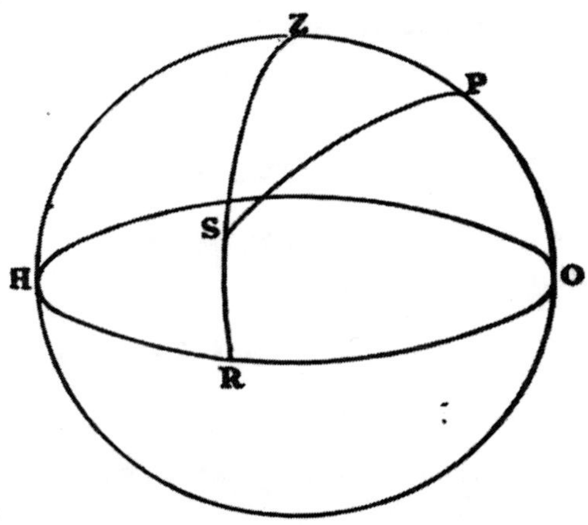

a vertical or azimuthal circle passing through the sun S; of which the angle S Z H = 45°, therefore S Z P = 135°, P S the arc of a declination circle = 90—*d*, and Z P = 90—*l*, *l* being the latitude;

R

Then in the spherical triangle P Z S we have, in circular functions,
$\text{Cos } S Z P \cdot \sin Z S \cdot \sin Z P = R^2 \cos S P - R \cos Z P \cdot \cos Z S. \ . \ . \ (1)$

[The Indian astronomy has two systems of trigonometry, one referring their problems to the trigonometry of the sphere, the other referring them to the right-angled triangle, of which the two sides are the Gnomon and its shadow, the third side being the corresponding hypotenuse.]

Assuming for these their initials g, s, h, equation (1) has to be transformed so that the sines and cosines may be expressed in terms of g, s, h and R, R being the assumed radius of the sphere.

In equation (1)

$$\text{Sin } Z P = \cos l, \quad \cos Z P = \sin l,$$

$$= g \cdot \frac{R}{h} \qquad = P \cdot \frac{R}{h},$$

P being the equinoctial shadow.

Also, $\cos S Z P = \cos 135°$, $\cos S P = \sin d$

$$= -\frac{R}{\sqrt{2}} \qquad = \frac{g}{h} \sin A,$$

A being the sun's rising amplitude.

Substituting these values respectively in equation (1), it becomes, when reduced,

$$-\frac{g}{\sqrt{2}} \cdot \sin Z = g \sin A - P \cos Z. \ . \ . \ . \ (2).$$

In which Z the zenith distance Z S is required.

Squaring equation (2), we have

$$\frac{g^2}{2} \cdot \sin^2 Z = g^2 \sin^2 A + P^2 \cos^2 Z - 2 g P \sin A \cdot \cos Z.$$

Or, since $\sin^2 Z = R^2 - \cos^2 Z$

$$g^2\left(\frac{R^2}{2} \sin^2 A\right) = \left(\frac{g^2}{2} + P^2\right) \cos^2 Z - 2 g P \sin A \cdot \cos Z.$$

Or, $$\frac{g^2\left(\frac{R^2}{2} - \sin^2 A\right)}{\frac{g^2}{2} + P^2} = \cos^2 Z - \frac{2 g \cdot P \sin A}{\frac{g^2}{2} + P^2} \cdot \cos Z.$$

$$\text{Now let Karani} = \frac{g^2\left(\frac{R^2}{2} - \sin^2 A\right)}{\frac{g^2}{2}+P^2} = \frac{144\left(\frac{R^2}{2} - \sin^2 A\right)}{72+P^2}, g \text{ being } 12$$

$$\text{And Phala} = \frac{g \text{ P} \sin A}{\frac{g^2}{2}+P^2} = \frac{12 \text{ P} \sin A}{72+P^2}$$

Then Karani $= \cos^2 Z - 2 \text{ Phala} \times \cos Z$, a quadratic for cos Z, consequently Karani+Phala^2 $= (\cos Z - \text{Phala})^2$, and the Kona-Sanku $= \cos Z = \sqrt{\text{Karani}+\text{Phala}^2} \pm \text{Phala} =$ sin sun's altitude, a solution identical with that given in rules from 28 to 32.

As a corollary the sine of the zenith distance, or

$$\text{Drig-jya} = \sqrt{R^2 - \text{Kona-Sanku}^2}.$$

Rule (33) then states that at the time the shadow of the Gnomon

$$S = R \frac{\text{Drig-jya}}{\text{Kona-Sanku}} = R \cdot \frac{\sin Z}{\sin a},$$

and the hypotenuse $= 12 \dfrac{\text{Drig-jya}}{\text{Kona-Sanku}} = 12 \cdot \dfrac{\sin Z}{\sin a}$.

Rules (34 to 36). Proceed to find the sun's altitude at any time from noon, when the hour angle H in degrees, the declination d, and the latitude l are given.

(34.) Assumes D the ascensional difference to be known, which from l and d can be easily computed.

And R$\pm$sin D is called the Antya.

It is the sine of the arc measuring the hour from sunrise till noon.

Then (Antya-Vers H)$\dfrac{\cos d}{R}$ is called Chheda $= (R \pm \sin D - \text{Vers H})$ $\dfrac{\cos d}{R}$ and Chheda $\times \dfrac{\cos l}{R} =$ Sanku or sin sun's altitude $= (R \pm \sin D - \text{Vers H}) \dfrac{\cos d \cos l}{R^2}$.

This result for finding the sun's altitude may be verified and explained as follows :—

In the adjoined figure let H Q Z P N represent the meridian, H E S N the horizon, s the place of the sun when rising; S, his place at any time from noon on the small diurnal circle s S; H the

a 2

corresponding degrees in the hour angle S P Z; Z the sun's zenith distance S Z; d the declination the complement of S P, and l the latitude $=90^\circ-$Z P.

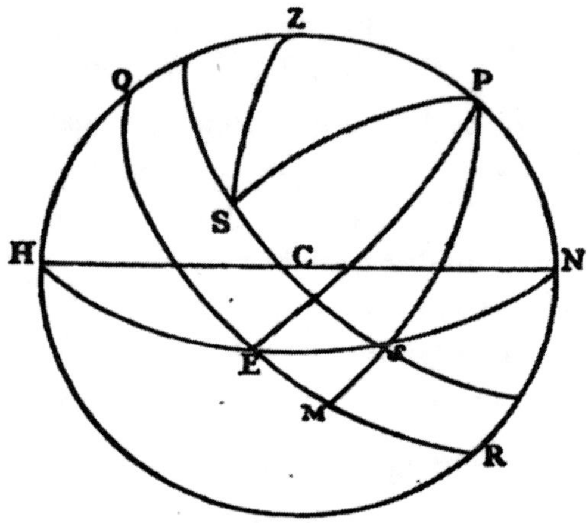

Hence in the spherical triangle S Z P,

$$\text{Cos H} = \frac{R^2 \cos Z}{\cos l \cos d} \mp R \tan l \cdot \tan d \quad . \ . \ . \ . \quad (1)$$

Also in the right angled spherical triangle E M S, in which E M $=$ D, the ascensional difference M S $= d$ and the angle, S E M $= 90^\circ - l$, and by Napier's rules

$$\sin D = \pm R \tan l \cdot \tan d. \quad . \ . \ . \ . \quad (2)$$

Therefore from (1) and (2) by addition

$$\text{Cos H} \pm \sin D) = \frac{R^2 \cos Z}{\cos l \cos d}$$

or $(\cos H \pm \sin D) \dfrac{\cos l \cos d}{R^2} = \cos Z =$ sin of the sun's altitude which is a result identical with that derived from Rules (34 to 36), since $\cos H = R - \text{vers H}$.

By way of corollary the Drig-jya or sin zenith distance is found $\sin Z = \sqrt{R^2 - \sin^2 a}$, a being the altitude.

Now by a reverse algebraic operation on the formula

$$(\text{Antya}-\text{vers H}) \frac{\cos d \cos l}{R^2} = \sin a$$

as derived in rules (34), (36), supposing the sun's altitude, with his declination and the latitude of the place to be known, any of the terms Antya, ascensional difference, the amplitude at rising, or the hour angle H may be found; such a method is adopted in Rules (37), (38) and (39) by reversing the calculation for the purpose of finding H, thus—

$$\frac{\text{R} \sin a}{\cos l} = \text{Chheda} = (\text{Antya} - \text{versed} \sin \text{H}) \frac{\cos d}{\text{R}}$$

$$\therefore \frac{\text{R}^2 \sin a}{\cos d \cos l} = \frac{\text{R Chheda}}{\cos d} = \text{Antya} - \text{vers H}$$

or

$$\text{Vers H} = \text{Antya} - \frac{\text{R}^2 \sin a}{\cos d \cos l}$$

$$\therefore \text{H} = \text{arc whose versed sine is Antya} - \frac{\text{R}^2 \sin a}{\cos d \cos l}$$

∴ H is found from the table of versed sines given in Chapter II., in which the radius is 3438, it is expressed in minutes of arc which are equivalent to pranas of time. For the sidereal day contains 21,600 pranas, and 360 degrees consists of the same number of minutes.

The calculation is made by successive steps, because there is less liability of error when a formula is taken arithmetically in parts, with distinct names, than as a whole in which the relation of the parts is not kept distinctly in view.

In our methods, complex formulæ are rendered more easy of solution by being in the first instance adapted to logarithmic computation, when traces of their origin are sometimes lost sight of.

A preparation is then made for calculating the sun's declination and his longitude at a given time; the latitude of the place, l, the sun's declination, d, the reduced amplitude A′, or the rising amplitude A, being also given.

From rule (40) we have—

$$\text{Sin } d = \frac{\cos l, \sin \text{A}',}{h_{,}} = \frac{\cos l, \sin \text{A}}{\text{R}}$$

$$\text{Also—Sin sun's longitude} = \frac{\text{R} \sin d}{\sin 24^{\circ}}$$

These are results derived from right angled spherical triangles, easily shown to be true by means of Napier's rules.

Rule (41) supposes that the end of the shadow of a vertical gnomon moves on the horizontal plane in the circumference of a circle. This is an error, which was refuted by Bhascara in the Goladhyhya. In fact, the locus of this point is a hyperbola.

Rules (42 to 44) would appear to need some preliminary explanation.

At any place on the equator the sphere of the heavens, as it appears to a spectator, is termed a right sphere.

If a projection of it on the meridian be represented by the figure E P Q *p*, in which the horizon is indicated by P *p*, the equator by E Q, and all the parallel diurnal circles, in which the heavenly bodies

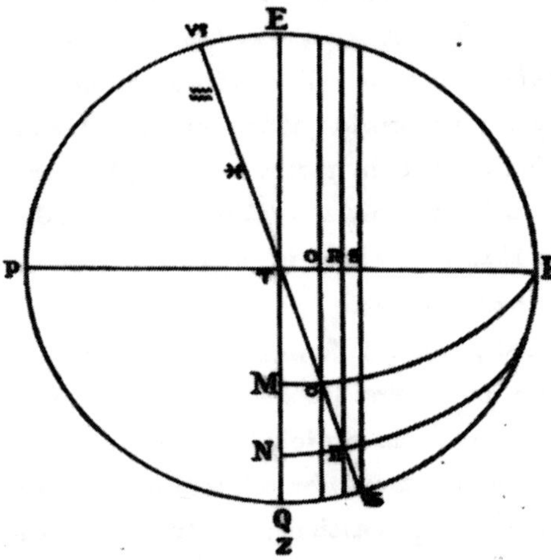

appear to move, will be projected in straight lines at right angles to the horizon. The first point of Aries will rise in the East, as at ♈ supposed in the figure, and the position of the ecliptic will then appear as a straight line ♈, ♉, ♊, etc.

The diurnal motion of the sphere will be round P *p* as an axis, the poles P, and *p*, being then in the horizon, and the first point of each sign will come to the horizon when rising, at points O R S. If great circles of the sphere be supposed to be drawn from the pole P through each of the points ♉ ♊ ♋, meeting the equator E ♈ Q in points M, N and Q, then the times taken by the three points of the ecliptic to rise at O, R and S, will be measured by the arcs of the equator ♈ M, ♈ N and ♈ Q, which are equivalent to the right ascensions of the three points, expressed in time; and ♈ M, M N and N Q will be the times taken successively for the rising of each sign and they are called the rising periods of the signs at the equator. The ascensions in a right sphere.

Rule (42) has for its object the determination of the right ascensions of the extremities of the first three signs at the ecliptic; it is expressed in words, equivalent to the formula

$$\text{Sin } \text{Æ} = \frac{\text{Cos } 24^\circ . \sin L}{\text{Cos } d}$$

To apply this rule to the calculation of the right ascensions, corresponding to longitudes of 30°, 60° and 90° of the ends of the three signs from the vernal equinox, the first step is to calculate the declination of each point by rule (40). In Hindu commentaries these are given respectively as

11° 43′, 20° 38′ and 24°.

For these arcs the cosines are then found to radius 3438′ and are 3366′, 3217′ and 3141′, which, with sines of the corresponding longitudes found from the table of sines, are substituted in the rule, or the above formula, by means of which the required right ascensions of the three ends are found to be

1670′, 3465′ and 5400′ or
27° 50′, 57° 45′ and 90°

The differences of the three right ascensions, namely,

1670′, 1795′, and 1935′, in arc

are equivalents of the same number of pranas reckoned in sidereal time.

They are the rising periods, or ascensions, of the first three signs, successively, in a right sphere.

The same numbers in a reverse order 1935, 1795 and 1670, are the rising periods of the next three signs, and the periods of the remaining six signs have the same values in the same order as those of the first six.

For all places on the earth, the right asceusions of the extremities of the signs have the same value as on the equator; but the rising periods of the signs, for places in north or south latitude, are the times of oblique ascension, in an oblique sphere; that is, in a sphere whose polar axis makes an angle with the horizon, equal to the latitude of the place.

The differences between the lengths of days and nights at places not on the equator, are, owing to the sun's apparent diural motions in the small circles of an oblique sphere.

At all places on the equator the days and nights are equal, although the arcs of 30° of each sign take different times in rising. For places between the equator and the arctic circle it is only when the sun is in either equinox, that the day between sun rise and sun set is equal to the night between sun set and sun rise. For other days, at places in north latitude, when the sun has a northern declination, the days are longer than the nights, but when the declination is south, the days are shorter than the nights; and in these cases, the difference in time between sun rise and six hours from noon is called the ascensional difference.

This difference in Hindu Astronomy is called Chara-Kala. The difference between the period of the rising of a sign in a given latitude and that of the same sign at the equator, is called the Chara-Khanda of that sign for the place.

For finding the rising periods of the first three signs at a given place, rule (43) continues by stating that, the ascensional differences of their ends are to be computed for the given place. If these be assumed respectively to be D_1, D_2 and D_3, then D_1, $D_2 - D_1$, and $D_3 - D_2$ are the Chara-Khandas at the place of the first three signs. These Chara-Khandas are then to be subtracted from the rising periods of the same three signs at the equator, and the remainders will be the rising periods in Pranas at the given place.

For the next three signs, their Chara-Khandas are added in a reverse order to the corresponding rising periods at the equator, and the sums will be the rising periods of these signs at the place.

The rising periods of the six signs thus found, taken in an inverse order, answer for the remaining six.

To make the subject more easily understood, let us assume the latitude of the place to be 22° 30″ north, then the ascensional differences D_1, D_2 and D_3 are 297′, 541′ and 642′.

The Chara-Khandas are, therefore, 297′, 244′ and 101′, either minutes of arc or pranas of time; and the accompanying table shows the rising periods of the twelve signs at the equator, and at places whose latitude is 22° 30′, together with the ascensional differences, respectively :—

	Times of rising at the equator in sidereal time.	Ascensional differences in sidereal time at places in north latitude 22½°.	Times of rising in sidereal time at places 22½° north latitude.
Aries	1670	− 297	1373
Taurus	1793	− 244	1549
Gemini	1937	− 101	1836
Cancer	1937	+ 101	2038
Leo	1793	+ 244	2037
Virgo	1670	+ 297	1967
Libra	1670	+ 297	1967
Scorpio	1793	+ 244	2037
Sagittarius ...	1937	+ 101	2038
Capricorn ...	1937	− 101	1836
Aquarius ...	1793	− 244	1549
Pisces	1670	− 297	1373

In the 13th Chapter of Surya Siddhanta, the point of the ecliptic just rising in the eastern horizon at any time is called the Udaya-

Lagna, or Horoscope; the point just setting is the Asta-Lagna; and the point on the meridian, the culminating point of the ecliptic, is called the Madhyama-Lagna.

The Udaya-Lagna is the point on which depends the casting of a nativity, or the construction of a scheme of the heavens, at the time of a birth.

It is of much importance for finding the Nonagesima point, and for other purposes in the Hindu method of calculating eclipses.

The operations on which rules (45), (46) and (47) of this, Third Chapter of the Surya Siddhanta, depend, are founded upon the foregoing rules relating to the rising periods of the signs, at a given place.

Rule (45). From the sun's longitude ascertained at a given time, find the Bhukta and Bhogya times in Pranas. Multiply the numbers of the Bhukta and Bhogya degrees (of the sign in which the sun is at the time) by the rising period of that sign, and divide the product by 30.

Rules (46), (47). From the given time in Pranas subtract the Bhogya time in Pranas, and the rising periods of the next signs (as long as possible, till a sign is arrived at whose rising period can no longer be subtracted; this sign is called the Asuddha sign or the sign incapable of subtraction). Multiply the remainder, that is found, by 30, and divide the product by the Asuddha rising period; add the quotient, in degrees, to the preceding signs reckoned from Aries. The result will be the place of the horoscope at the eastern horizon.

If the time at the end of which the horoscope is to be found be given before sunrise, then take the Bhukta time and the rising periods of the signs preceding that which is occupied by the sun, in a contrary order from the given time.

Multiply the remainder by 30, and divide the product by the Asuddha rising period. Subtract the quotient in degrees from the signs; the remainder will be the place of the horoscope at the eastern horizon.

The following is an example of the method of calculating the horoscope in accordance with these rules:—

Suppose the latitude of the place to be 22° 30′, which is about 5′ south of the ancient city of Dhar, in Malwa, and 41′ south of Ojein, for one of which places the table of the risings of the signs given above may have been intended; and let the sun's place at a given time, say 5 hours 15 minutes reckoned from sunrise, be eight signs 20°, supposed to be calculated from tables, or rules of the Second Chapter for the given day and hour.

Then, his place would be 20° in the sign Sagittarius, and would divide that sign in the proportion of two to one. These parts are named in the rules the Bhukta and Bhogya degrees, and the rising period of Sagittarius, from the table is 2033 Pranas, which, divided, in the proportion of two to one, gives 1,358⅔ Pranas and 679⅓ Pranas the Bhukta and Bhogya times.

Now the given time from sunrise being 5 hours 15 minutes, or 18,900 seconds, and expressed in Indian form, we have—

		Pranas.
The given time from sunrise	=	4,725
Subtracting from this the Bhogya time	=	679⅓
The remainder becomes the time when the first Capricorn rose	=	4,046⅔
Again, subtracting from this remainder the rising period of Capricorn	=	1,836
We have time since the rising of first Aquarius ..	=	2,109⅔
Then subtracting the rising period of Aquarius.. ..	=	1,549
Or the time since the first of Pisces rose	=	560⅔

But the rising period of Pisces cannot be subtracted from the above remainder; Pisces, therefore, in this case has the name of the Asuddha sign.

To determine the proportional part of the sign itself, we have—

Rising period of Pisces : 560⅔ : : 30° : proportional part.

$$\therefore \text{ proportional part of Pisces above the horizon} = \frac{560\frac{2}{3}}{1373} \times 30$$

Or the Lagna is = 12° 15′ 1″ from the beginning of Pisces, or 17° 45′ from the Equinox.

The calculation from the Lagna, or Horoscope, was of great importance in the Hindu theory of a solar eclipse; it was used also in the rules of computing the conjunctions of planets, and in those of their heliacal risings and settings. It is also obvious that the nonagesimal point of the ecliptic, which is at a distance of 90°, measured on that circle from the point of it which is the momentary horoscope, is at once found when the Lagna is known. The Azimuth of the nonagesimal point, is likewise found at once from the amplitude of the Lagna, by the addition or subtraction of 90° measured on the horizon, all involving less labour than the more complex rules given in our works on Astronomy of about 150 years ago.

The time being given, as assumed in the preceding rules, and the place of the sun being found for that time, rule (48) indicates the method of finding the Madhya-Lagna, or the point of the ecliptic then on the meridian, *i.e.*, the point commonly called the culminating point of the ecliptic.

First, the hour angle from noon is to be found and its equivalent in Pranas of equatorial time; and the rising periods of the signs, with their Bhukta and Bhogya time, corresponding to this horary angle, are to be estimated by a method similar to that employed for the horoscope. Then the arc of the ecliptic in signs, degrees, etc., indicated by this estimated time, is to be added to the place of the sun, or subtracted from it, as the case may be, for times before or after noon. The result of this process gives the place of the culminating point of the ecliptic.

Rule (49) is a converse rule to that for finding the Lagna. The object is to find the time from sunrise, when the place of the horoscope and that of the sun are assumed to be known.

The text of the rule is: "Find the Bhogya time in Pranas of the less (longitude), and the Bhukta times of the greater, add together these times and the rising periods of the intermediate signs (*i.e.*, between the two given longitudes, or places of the sun and the horoscope) and you will find the time."

Rule (50) states the various cases that may occur in the foregoing rule.

"When the given place of the horoscope is less than that of the sun, the time will be before sunrise; but when it is greater, the time will be after sunrise. And when the given place of the horoscope is greater than that of the sun, increased by six signs, the time found from the place of the horoscope and that of the sun added to six signs, will be after sunset."

Rule (50) which determines the time for sunrise (when the place of the sun, and that of any point of the ecliptic just rising on the eastern horizon, are both given), reckoned from either equinox, is in a great measure applicable to the risings of the five planets, whose latitudes are generally small, and which may have their places at any degree of longitude.

CHAPTER IV.

ON THE HINDU METHOD OF CALCULATING THE OCCURRENCE OF THE ECLIPSES OF THE MOON.

The day on which a Lunar Eclipse will happen is to be found by comparing the places (or longitudes) of the moon and her node on the day of the moon's opposition with the sun, when it is presumed the eclipse will take place, and if at the moment of the opposition the difference of the longitudes of the moon and her node be within about $7\frac{1}{2}$ degrees, there will be an eclipse.

In Chapter IV. the sun's mean diameter is assumed = 6,500 Yojanas, and the moon's mean diameter is assumed = 480 Yojanas.

On account of the variable distances of the sun and the moon, their apparent diameters are greater when near than when more remote, and a correction is applied on the hypothesis that the apparent magnitudes vary with the daily motions, which also are in the inverse ratio of the distances.

The mean daily motions of the sun and the moon are found by dividing the revolutions made by each in a Maha-Yuga by the number of days in the same Yuga, taken from Table I. of Chapter I. of the Siddhanta.

Thus, the mean daily motion of the sun $= \frac{4,320,000}{1,577,917,828}$, this reduced to minutes = 59·13616′, and

$$\text{The mean daily motion of the moon} = \frac{57,753,336}{1,577,917,828}.$$

The daily motions of the sun and moon on the day of the eclipse are called their true daily motions.

Rule (2) is that "The diameters of the sun and moon multiplied by their true diurnal motions, and divided by the mean diurnal motions, become the Sphuta or rectified diameters."

If σ and μ be taken to denote the true diurnal motions of the sun and moon in minutes on the day of the eclipse, then

$$\text{The sun's rectified diameter} = \frac{6500 \times \sigma}{59 \cdot 13616}, \text{ and}$$

$$\text{The moon's rectified diameter} = \frac{480 \times \mu}{790 \cdot 56}.$$

Rule (3). "The rectified diameter of the sun multiplied by his revolutions (in a Maha-Yuga) and divided by the moon's revolutions in that Yuga, or multiplied by the periphery of the moon's orbit and divided by that of the sun, becomes the diameter of the sun at the moon's orbit."

Hence, after reduction of the large numbers here employed,

$$\text{The diameter of the sun at the moon's orbit} = \frac{6500 \times \sigma}{790 \cdot 56} \text{ Yojanas}$$
$$= 8 \cdot 222 \times \sigma.$$

The circumference of the moon's orbit is reckoned to be 324,000 Yojanas, and the number of minutes of arc in the same circumference being 21,600. Therefore, 15 Yojanas correspond with one minute of arc, and the above diameter of the sun, divided by 15, gives:
The apparent diameter of the sun's disc in minutes of arc

$$= \cdot 54813 \times \sigma$$

$\therefore$ The mean apparent diameter of the sun's disc

$$= \cdot 548 \times 3 \times 59 \cdot 13616 = 32 \cdot 3943' \text{ nearly.}$$

The rectified diameter of the moon, divided by 15, gives:
The apparent diameter of the moon's disc in minutes

$$= \frac{480 \times \mu}{790 \cdot 56 \times 15} \text{ nearly} = \cdot 04048 \times \mu.$$

$$\left.\begin{array}{r}\text{And the mean apparent diameter of the disc} \\ \text{of the moon}\end{array}\right\} = 32 \text{ minutes.}$$

For the diameter of a section of the earth's shadow at the moon is found by rules (4) and (5). "Multiply the true diurnal motion of the moon by the earth's diameter, and divide the product by her mean diurnal motion; the quantity obtained is called the Suchi."

The earth's diameter is estimated to be 1600 Yojanas.

$$\therefore \text{ The Suchi} = \frac{1600 \times \mu}{790 \cdot 56} \cdot \text{Yojanas} = 2 \cdot 024 \times \mu \text{ nearly.}$$

The calculation then proceeds, in rule (4): "Multiply the difference between the earth's diameter and the rectified diameter of the sun, by the mean diameter of the moon, and divide the product by that of the sun."

The operation is indicated by—

$$\left\{\frac{6500 \times \sigma}{59{\cdot}13616} - 1600\right\}\frac{480}{6500} \text{ Yojanas.}$$

This amount is then to be subtracted from the Suchi, and the remainder is the earth's shadow at the moon in Yojanas

$$= \frac{1600 \times \mu}{790{\cdot}56} - \left\{\frac{6500 \times \sigma}{59{\cdot}13616} - 1600\right\}\frac{480}{6500} \text{ Yojanas,}$$

and dividing by 15 to convert the Yojanas to minutes:

The diameter of the earth's shadow at the moon in minutes of Arc,

$$\text{becomes} = 106\tfrac{2}{3} \times \frac{\mu}{790{\cdot}56} - 32 \times \frac{\sigma}{59{\cdot}136} + 7\tfrac{7}{8}.$$

If we make $\mu = 790{\cdot}56'$ and $\sigma = 59{\cdot}136'$, the mean motions of the sun and moon.

The mean diameter of the earth's shadow reduces to

$$106\tfrac{2}{15} + 7\tfrac{7}{8} - 32 = 82 \text{ minutes nearly.}$$

Rule (6). "The earth's shadow is always six signs from the sun. When the place of the moon's node is equal to that of the shadow, there will be an eclipse, or, when the node is some degrees within, or beyond, the place of the shadow, the same thing will take place."

Rules (7), (8). The longitudes of the sun and moon being computed for the midnight preceding, or after conjuncton or opposition, proportional parts are to be applied for the changes of their places in the interval between.

Rule (9). "The moon being like a cloud in a lower sphere, covers the sun in a solar eclipse; but in a lunar eclipse the moon moving eastward enters the earth's shadow, and the shadow obscures her disc."

To find the magnitude of an eclipse: Let D be the diameter of the coverer, d the diameter of the body eclipsed, λ the latitude of the moon at the time of Syzygy.

Rules (10), (11). The quantity of the eclipsed part of the di- will be $= \frac{1}{2}(D \pm d) - \lambda$.

If this quantity be greater than the diameter of the disc of the body undergoing eclipse, the eclipse will be total; otherwise, it will only be partial.

But there will be no eclipse if λ is greater than $\frac{D + d}{2}$.

Rule (12.) "Find the halves, separately, of the sum and difference of the diameters of that which is to be covered and that which is the coverer..

"Subtract the square of the moon's latitude from the squares of the half sum and the half difference and take the square roots of the results."

Rule (13.) "These roots, multiplied by 60 and divided by the diurnal motion of the moon from the sun, give the Sthity-ardha, the half duration of the eclipse and the Mard-ardha, the half duration of the total darkness, in Ghatikas (respectively)."

If these be denoted by S and M and the daily separation of the moon from the sun by l,

$$\text{Then } S = \frac{60}{l} \times \sqrt{\left(\frac{D+d}{2}\right)^2 - \lambda^2} \text{ and}$$

$$M = \frac{60}{l} \times \sqrt{\left(\frac{D-d}{2}\right)^2 - \lambda^2}$$

To illustrate the method of calculation by a figure :—

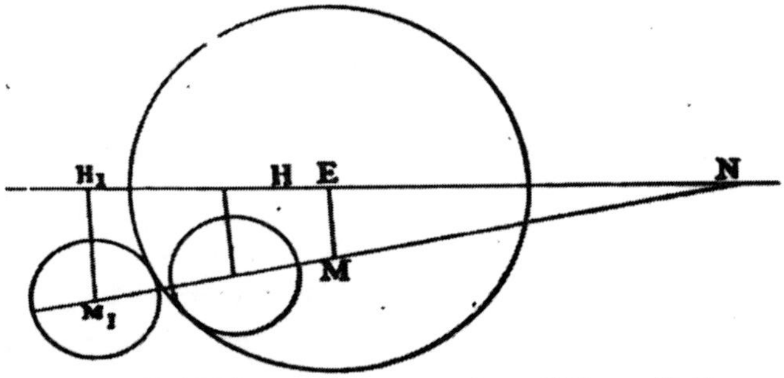

Let the line H_1 E N represent a portion of the ecliptic, and M_1 M N a part of the moon's path interesting the ecliptic in the ascending node N.

8

If E and M be supposed to be the centres of the earth's shadow, and of the moon, at the instant of opposition, that is, at the time of the full moon, then E M will represent the latitude of the moon, at that time (which may be denoted by T), and E M will be $= \lambda$.

Again, let H and M_1 be the places respectively of the centres of the shadow, and of the moon, at the beginning of the eclipse, or the moment of the first contact of the moon with the shadow, then H_1 H is the difference of the moon's longitude from her place at the first contact, and her place at full moon.

The arc H_1 H is found approximately by assuming the moon's latitude, λ, to remain for a short time unchanged and that in the triangle M_1 M E

$$M_1 E^2 = E M^2 + M M_1{}^2,$$

But $M_1 E = \frac{D + d}{2}$, $E M = \lambda$ and $M_1 M = H_1 H$ nearly,

or $H H_1 = \sqrt{\left(\frac{D + d}{2}\right)^2 - \lambda^2}$, but l being the assumed daily relative motion of the sun and the moon and S, the half duration of the eclipse

$$\frac{S_1}{60} = \frac{H H_1}{l}$$

$$\text{or } S = \frac{60}{l} \times \sqrt{\left(\frac{D + d}{2}\right)^2 - \lambda^2}$$

From the daily motions of the sun, the moon and the node, proportional parts of their longitudes are to be computed for changes in them, during the time S; these are to be applied by subtraction from the places found for the sun and moon at the time of the opposition, but by addition to the place of the node at that time.

Then by means of the corrected places of the moon and her node, the moon's latitude is to be computed, and this being substituted in the above formula a nearer approximation is obtained for S.

The process is to be repeated until the value obtained for S is the same in each repetition.

This value of S is called the exact first Sthity-ardha.

To find the second Sthity-ardha, or that for the end of the eclipse, the proportional changes in the places of the sun and moon are now to be added to their place at the opposition, but the change in the place of the moon's node is to be added to the place at the opposition.

From these corrected places, the moon's latitude is again to be computed and substituted for λ in the above formula, for a nearer value of S, at the last contact.

The same process is to be repeated until the exact second Sthity-ardha is found.

In like manner, the first and second Mard-ardhas are determined by repeated calculations.

Rule (16). The middle of the lunar eclipse is reckoned to occur at the time of the full moon.

If this time be denoted by T, then

T – 1st S is the time of the first contact with the shadow and T + 2nd S is the time of the end of the eclipse, also

(17). T – 1st M and T + 2nd M are the times of the beginning and end of the total darkness.

To determine the amount of obscuration at a given time during the continuance of an eclipse :—

The quantity of the eclipsed part gradually increases to the middle of the eclipse, and it is determined at any moment, by the time elapsed from the beginning, or first contact, which may here be denoted by m.

A proportional part of the variation in longitude is to be computed in minutes of arc for the time $S-m$, S as before being the 1st Sthity-ardha, or half the duration.

If the relative daily motion in longitude be denoted by l, the difference in longitude at the moment, from that at the middle of the eclipse would be in minutes of arc

$$= \frac{l}{60}(S-m)$$

s 2

This difference is called the Koti; the perpendicular of a right angled triangle of which the base is the moon's latitude, and the hypothenuse is the distance in arc between the centres of the moon and the earths shadow in the lunar eclipse, and between the centres of the moon and the sun in the case of a solar eclipse.

$$\text{The eclipsed part in minutes} = \frac{D+d}{2} - \sqrt{\text{Koti}^2 + \lambda^2}$$

Rule (21). A similar method is employed for calculating the eclipsed part at a given time between the middle of the eclipse and the end, in which case the second Sthity-ardha is used for finding the Koti or perpendicular of the right angled triangle.

Rules (22—23). In these rules the converse of the above proposition is propounded.

The quantity of the eclipsed part is supposed to be given in minutes of arc, for which the corresponding time in Ghatikas is to be found by a method similar to that in rule (13); the process being repeated when a nearer approximation is desired.

If n denotes the minutes of arc of the eclipsed part of a lunar eclipse, then

$$\text{Koti} = \sqrt{\left(\frac{D+d}{2} - n\right)^2 - \lambda^2}$$

and in a solar eclipse

$$\text{The Koti} = \frac{\text{Apparent Sthity}}{\text{Mean Sthity}} \times \sqrt{\left(\frac{D+d}{2} - n\right)^2 - \lambda^2}$$

From the Koti the time is found in Ghaticas, as in the method of finding the Sthity-ardha.

ON THE VALANAS.

It is remarked in the Surya Siddhanta that the phases of an eclipse cannot be exactly understood without their projection, and the Hindu method of projection is explained in Chapter VI.

Here, however, two rules (24 and 25) are given for finding what are termed the Valanas, two angles whose sum or difference constitutes the so-called rectified Valana, or "variation of the ecliptic."

As an entire variation, it is equal to the angle between a circle of latitude through the place of a body on the ecliptic, and the circle of position through the same place; the circle of position being defined as the great circle, passing through a planet, and through the north and south points of the horizon.

To find the Valanas by

"Rules (24 and 25). Find the zenith distance of the circle of position passing through the body, multiply its sine by the sine of the latitude of the place, and divide the product by the radius. Find the arc whose sine is equal to the quotient; the degrees contained in this arc are called the degrees of the Aksha, or latitudinal Valana; they are north or south, according as the body is in the eastern or western hemisphere of the place.

(25.) "From the place of the body, increased by three signs, find the variation (which is called Ayana or solstitial Valana). Find the sum or difference of the degrees of this variation and those of the latitudinal Valana, when those are of the same name or of contrary names; the result is called the Sphuta or true Valana.

"The sine of the true Valana, divided by 70, gives the Valana in digits."

In order to explain the above two rules, let R Z P N represent the

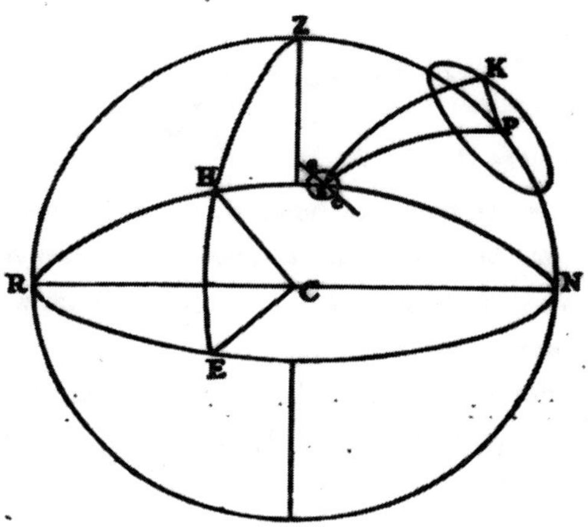

meridian of the place, Z the zenith, P the pole of the equinoctial, N the north point of the horizon R E N, Z H E the prime vertical.

Also suppose S to be the place of the body to be eclipsed, and through S the circle of position N S R to be drawn passing through the north and south points N and R of the horizon.

Then the object of the Valana is to determine the position of a short arc of the ecliptic ϵ S *c*, as it would appear to an observer at a given place, if visibly traced on the disc of the sun or the moon.

Now if K be the position of the pole of the ecliptic at the time of the eclipse, and P the pole of the equinoctial, it is obvious that the small arc ϵ S *c* will be at right angles to S K, the circle of latitude through S; and the rectified Valana would be the angle K S N between the circle of latitude S K and the circle of position S N, or the angle K S N.

But in many respects, it is more convenient to calculate separately, the two angles composing it, namely, P S N, the Aksha Valana, and K S P called the Ayana Valana, and from these to form the true Valana by addition or subtraction as may be found necessary.

First. In the spherical triangle P S N

$\text{Sin } PSN = \dfrac{\sin PNS \cdot \sin PN}{\sin SP}$, but angle P N S is measured by Z H = N suppose, P N = the latitude = l and S P = the complement of the declination,

$$\therefore \quad \left.\begin{array}{r}\text{Sine of the Aksha}\\ \text{or latitudinal Valana}\end{array}\right\} = \frac{\sin N \,.\, \sin l}{\cos d}$$

In the text we have the Sin Aksha $= \dfrac{\sin N \sin l}{R}$,

In which R here is to be understood, as the radius of the sun's diurnal path, on the day of the eclipse, and consequently, the sun having a supposed declination d, the cosine of the arc d would be the radius of the diurnal circle.

Secondly. For the Ayana Valana the rule only directs it to be found from the place of the body increased by three signs, or longitude L + 90°.

In the spherical triangle P S K.

P K the measure of the obliquity is reckoned to be 24°, P S is the co-declination, and the angle S K P = L + 90°.

∴ We have in the spherical triangle K S P, the sine of the angle K S P, or Sine Ayana or Solstitial Valana $\Big\} = \dfrac{\sin(90 + L) \cdot \sin 24°}{\cos d}$

In which cos *d*, as before, is represented in the text by R, the radius of the diurnal circle, whose declination is the arc *d*.

The angle called the Ayana, is obviously the same as that which is called by astronomers the angle of position.

Thus, it will be seen that the rules (24) and (25) deal only with the ascertaining of the angles known as the Valana, which angles give means of projecting the line of the ecliptic upon the disc of the body eclipsed.

CHAPTER V.

ON THE CALCULATION OF A SOLAR ECLIPSE.

It has been seen already (at the end of the description of the third Chapter), how the Hindus by means of the rising signs, determined the place of the horoscope or the point of the Ecliptic just rising, at any time, in the Eastern horizon—the point called by the Hindus the Udaya Lagna—and how, by similar means, they found the culminating point of the Ecliptic.

To reckon 90° along the Ecliptic, from the point of it just rising, became also an easy method of finding the point which among modern astronomers goes by the name of the nonagesimal.

This point on the occasion of a solar eclipse was of importance in its connection with parallax.

Verse I., Chapter V., begins by stating that there is no parallax in longitude, when the sun's place is equal to the place of the nonagesimal, and that when the north latitude of the place is equal to the north declination of the nonagesimal point (that is when the nonagesimal point is in the zenith of the place) there will be no parallax in latitude.

Rules are then laid down, as a preparation for calculating the parallaxes both of the latitude and the longitude when the place of a planet has different positions, *i.e.*, when the sun is to the east or west of the nonagesimal.

By rule (3) the amplitude of the horoscope is determined.

The place of the horoscope, at the instant of the conjunction reckoned from sunrise, is to be found by means of the rising periods.

The sine of the longitude of this point is then multiplied by the sine of 24°, the sun's greatest declination, and the product divided by the cosine of the latitude of the place.

The result is the Udaya, or the sine of the amplitude of the horoscope, thus—

$$\text{Udaya} = \frac{\text{sine L} \cdot \sin 24^\circ}{\cos l}$$

L being the longitude of the Lagna or horoscope, and l the latitude of the place.

(4). The place of the culminating point of the ecliptic is then to be found by means of the rising periods of the signs, and from the longitude of that point its declination is to be calculated; let it be denoted by d, and the latitude of the place by l. Then $l \pm d$ is the meridian zenith distance of the culminating or middle point.

(5) and the sine of the zenith distance $\sin(l \pm d)$ is called the Madhyajaya, the sine of the middle point.

To illustrate some of the terms here used in the rules: Let R M Z P O represent the meridian; R E O the projection of the horizon, Q E C that of the equator, E the east point, Z the zenith and P the pole of the equinoctial.

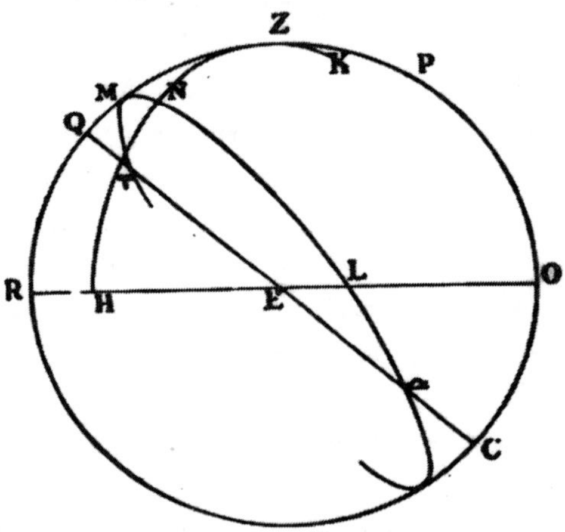

Also let ♈ M N L ♎ represent the ecliptic, M its culminating or middle point, L its lagna or rising point, N the point nearest the zenith or the nonagesimal, K the pole of the ecliptic, and H N Z K the vertical circle passing through N.

Hence ♈ being assumed as the vernal equinox, ♈ M will represent the longitude of the culminating point, ♈ N that of the nonaagesimal and ♈ L of the horoscope or rising point.

It is obvious from the figure since L Z, L N and therefore also L H represent quadrants that L H is equal to E R which is also a quadrant, and if from each of these equals the common arc H E be taken, the remaining arcs R H and E L are equal.

But E L represents the amplitude of the rising point, the sine of which, or the Udaya is found by rule (3), and the arc R H measures the angle R Z H, or M Z N.

Now in the right angled spherical triangle M Z N,

$$\text{Sin M N} = \frac{\text{Sin M Z}}{\text{R}} \text{ sin M Z N}$$

or substituting from rules (3) and (5)

$$\text{Sin M N} = \frac{\text{Madhyajaya} \times \text{Udaya}}{\text{R}}$$

$$= \frac{\sin (l \pm d) \times \sin 24^\circ . \sin \text{L}}{\text{R} \cos l}$$

Rule (5). The zenith distance N Z and the altitude N H of the nonagesimal point, are found approximately from their sines, that of the zenith distance of N being called the Drikshepa, that of the altitude Driggati.

To find the Drikshepa, "Multiply the Madhyajya by the Udaya, divide the product by the radius and square the quotient. Subtract the square from the square of the Madhyajya; the square root of the remainder is (nearly equal to) the Drikshepa, or the sine of the zenith distance of the nonagesimal, or the sine of the latitude of the zenith."

For the Driggati, "The square root of the difference between the squares of the Drikshepa and the radius is the sine of the altitude of the nonagesimal.

"The sine and cosine of the zenith distance of the culminating point are reckoned the rough Drikshepa and Driggati respectively."

The rule for the zenith distance of the nonagesimal is obviously derived from the right angled spherical triangle M Z N of the figure, by considering the sines of its sides as if they were sides of a plane right angled triangle, thus

$$\text{Sin } Z\,N = \sqrt{\sin^2 Z\,M - \sin^2 M\,N}$$

In which sines of Z M and M N have been detailed above.

PARALLAX.

The moon's parallax in longitude, on the occasion of a Solar Eclipse, involves a series of complex calculations, which for convenience, are divided into steps.

The true time of conjunction of the sun and moon differs from the apparent time by the relative parallax of the sun and the moon expressed as time.

Hindu astronomers estimate the moon's horizontal parallax to be $\frac{1}{15}$ of the mean daily motion in her orbit.

But the moon's daily motion is 13° 10′ 46·7, which divided by 15 gives 52′ 42″ as her horizontal parallax.

On the same hypothesis they reckoned the sun's horizontal parallax to be 3′ 56″ and the relative horizontal parallax to be 48′ 46″.

The equivalent of this in time was estimated to be 4 Ghatikas, the fifteenth part of a day.

Rule (7). The first step is to compute a divisor called the Chheda

$$= \frac{(\sin 30)^2}{\text{Driggati}} = \frac{R^2}{4 \text{ sine altitude of nonagesimal}}$$

If the difference of longitudes of the nonagesimal and of the sun be denoted by D, then rule (8), the moon's parallax in longitude from the sun, expressed in Ghatikas,

$$= \frac{D}{\text{Chheda}}.$$

This will be a first approximation to the relative parallax in time, and the continuation of the process will be understood from the text.

Rule (9). "Subtract the parallax in time (just found) from the end of the true time of conjunction, if the place of the sun be beyond that of the nonagesimal; but if it be within add the parallax.

"At the applied time of conjunction, find again the parallax in time, and with it apply the end of the true time of conjunction, and repeat the same process of calculation until you have the same parallax, and the applied time of conjunction in every repetition. The parallax lastly found is the exact parallax in time and the time of the conjunction is the middle of the solar eclipse."

The relative parallax in latitude of the moon from the sun is found from rule (10). Multiply the Drikshepa (sine of zenith distance of nonagesimal) by the relative daily motion of the sun and moon, and divide the product by 15 times the radius. Thus,

Relative parallax in latitude $= \frac{48\frac{1}{5}}{3438} \times$ sin zenith distance of nonagesimal, or (11),

$$\text{Parallax in latitude} = \frac{\text{Drikshepa}}{70} = \frac{\text{sin zenith distance of nonagesimal}}{70}$$

Rule (12). "The amount of the parallax found is north or south, according as the nonagesimal is north or south of the zenith. Add the amount to the moon's latitude, if they are of the same name; but, if of contrary names, subtract it. (The result is the apparent latitude of the moon.)"

The apparent time of conjunction having been found, by applying the parallax in longitude, expressed as time, to the computed true time of conjunction, as indicated in rule (9); and for this apparent time the moon's apparent latitude having been calculated, according to rule (10), by applying the parallax in latitude to the true latitude, the method of procedure afterwards differs little from that employed in Chapter IV. on lunar eclipses.

Rule (13). "In the solar eclipse, with the apparent latitude of the moon, find the Sthity-ardha (or half duration) the Mard-ardha (or half the total darkness), etc., of the eclipse, as before mentioned; also the Valana (or deviation of the ecliptic), the eclipsed portions of the disc at assigned times, etc."

The first approximations to the times of beginning and ending, etc., having been computed, the process of finding the effects of parallax is renewed where necessary for each of such times, as detailed in rules (14, 15, 16 and 17). "Find the parallaxes in longitude (converted into time) by repeated calculation at the beginning of the eclipse, found by subtracting the first Sthity-ardha (just found) from the time of conjunction, and at the end, found by adding the second Sthity-ardha.

"If the sun be east of the nonagesimal, and the parallax at the beginning be greater, and that at the end be less than that at the middle; or if the sun be west, and the parallax at the beginning be less, and that at the end be greater than the parallax at the middle, add the difference between the parallaxes at the beginning and middle, or at the end and the middle to the first or the second Sthity-ardha (above found); otherwise, subtract the difference.

"It is then when the sun is east or west of the nonagesimal at the times both of the beginning and the middle, or of the middle and the end, otherwise add the sum of the parallaxes (at the time of the beginning and middle, or of the end and the middle) to the first or the second Sthity-ardha.

"(Thus you have the apparent Sthity-ardhas, and from these the times of the beginning and the end of the eclipses of the sun.)

"In the same manner find the apparent Mard-ardha (and the times of the beginning and end of the total darkness in the total eclipses of the sun)."

CHAPTER VI.

ON THE PROJECTION OF SOLAR AND LUNAR ECLIPSES.

The object of a projection is to shew, by a figure, the points on the disc of the body to be eclipsed at which the obscuration begins or ends, &c.

In the beginning of the VI. Chapter of the Surya Siddhanta, it is stated that the phases of an eclipse cannot be exactly understood without a knowledge of their projection.

In a lunar eclipse, the moon's eastern side becomes first immersed in the shadow, and the western side is the part that emerges.

In a solar eclipse, the western side of the sun's disc is first obscured, and the eastern side is the part last relieved from the body of the moon.

It is of importance in a projection to know the position of the line which would represent on the disc of the body to be eclipsed, the apparent direction of the ecliptic, or the direction in which the sun is moving.

This direction is fixed with reference to the place of an observer, by means of the rectified or true Valana deduced in rule (25), Chapter IV.

The circle in which the Valana is to be marked is thus described:

Rule (2). "Having marked at first a point on the (chunam) floor, levelled with water, describe on the point as centre, a circle with radius equal to 49 digits."

The scale of projection is thus the same as that of the Gnomon, of 12 angulas or digits, in which the shadows cast by rays from the sun and moon, and the fainter rays from the other celestial bodies were estimated.

The radius assumed in the first circle is a little over four times the ordinary Gnomon, of 12 digits. It is connected with the hypothetical Hindu radius 3,438 minutes of arc, by supposing the two radii to be equal; consequently, the digit adopted would be equal to nearly $70\frac{1}{3}$ minutes of arc, of the same circle. It was assumed to be 70 integral minutes.

The elements of an eclipse such as the moon's latitude, diameter, the Valana, eclipsed parts, &c., which were expressed in minutes of arc, were reduced and converted into digits, when desirable, by simply dividing the minutes by 70.

But angulas or digits are of uncertain magnitude, they are of various dimensions in different books.

If we assume the ordinary digit to have been about three quarters of an inch, the radius of the first circle would have been about 37 inches.

Rule (3) directs a second circle to be described on the same centre with a radius equal to half the sum of the coverer and the covered and also a third circle with a radius equal to the semi-diameter of that which is to be covered.

In a lunar eclipse, the coverer is the earth's shadow, the diameter at a mean distance of the moon subtending an angle estimated at about 82 minutes, and the body to be covered, the moon, the diameter of whose disc at the mean distance was estimated to subtend an angle of about 32 minutes, or

$$\frac{D + d}{2} = 57'$$

$$\text{and} \quad \frac{d}{2} = 16'$$

Thus, if the radii were taken on the same scale with the radius of the first circle, the radius of the second circle, would have been only ·6 of an inch and that of the third about $\frac{2}{3}$ of a digit or $\frac{1}{4}$ of an inch.

It is therefore obvious that for practical purposes the radii of the second and third circles must have been drawn on a different scale from that of the first, and that the first circle was merely used for laying down the angle of the Valana, or the angle which the direction of the ecliptic made with the east and west line of the projection.

The description of the method of projection in general terms, must necessarily be defective; for projection cannot dispense with computations and these imply numerical data for the day on which an eclipse is expected to take place.

For example, the longitudes of the sun, the moon and moon's nodes, and their true daily motions have to be ascertained for the time, the latitudes of the moon for the computed times of the beginning, middle and end of the eclipse, quantities which change in value and position by the progress of the moon in which it may cross the ecliptic during the obscuration.

If the three dark lined circles in the adjoined figure be supposed

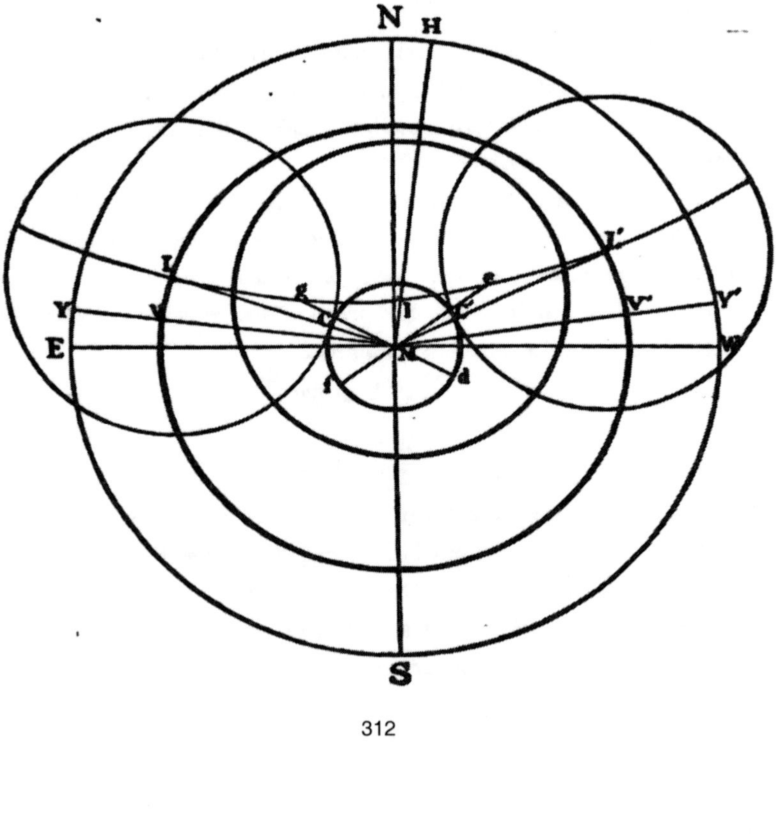

to represent those projected in accordance with rules (2) and (3) having M as a common centre, which is here assumed to be the centre of the moon, whose disc is represented by the third circle.

N S and E W are drawn as north and south and east and west lines, as mentioned in beginning of the third chapter.

Then, two lines M V and M V′ are drawn, making angles E M V and W M V′ equal to the computed angle named the rectified Valana. These lines represent the position of the ecliptic at the beginning and end of the eclipse. Here they intersect the second circle in the two points V and V′.

The moon's latitude is to be found for the computed beginning of the eclipse, or first contact of the moon's disc with the earth's shadow, and a perpendicular L V is to be drawn from L in the second circle equal to the minutes of arc in the sine of the moon's latitude.

If now from L as a centre with a radius equal to the minutes in the semi-diameter of the earth's shadow a circle be described, it will touch the third circle on the moon's disc in some point C′ which will be the point of first contact. If in like manner for the computed end of the eclipse, the moon's latitude again be found, and laid by means of the minutes in its sine, as a perpendicular L′.V′ from L in the second circle. then from L′ as a centre with a radius equal to the semi-diameter of the earth's shadow, if a circle be described it will touch the circle representing the moon's disc in some point C′, which will be the point of last contact, at the end of the eclipse.

For the middle of the eclipse at the time of the opposition, the Valana is to be marked from near one of the ends of the north and south line; there is considerable obscurity in the directions for drawing the line making an angle with N S equal to the Valana. This line when its position has been correctly drawn, is here represented by H M and supposed to be at right angles to the position of the ecliptic at the instant of the full moon. The moon's latitude is found and laid upon this line (as I M suppose), then I, a point on the ecliptic, will be the place of the earth's shadow at the time; and if

T

from this point as a centre a circle be described with a radius equal to the semi-diameter of the earth's shadow, the part of the moon's disc covered by it will be the eclipsed part which may be partial or total.

If the moon's disc be conceived to be fixed in the projection, the relative path of the earth's shadow is found by describing the arc of a circle through the three points L I L′ here understood to be projected points on the ecliptic, and by assuming any other intermediate point of this arc as a centre, and the semi-diameter of the shadow as a radius, the circle that would be described would cover a portion of the moon's disc, which would represent the magnitude of the eclipsed part corresponding to that point in the progress of the eclipse.

In a total lunar eclipse, the point of the moon's disc at which the total darkness begins is to be found by drawing a line from the common centre M, of a length equal to half the difference of the diameter of the earth's shadow and of the moon or $\frac{D-d}{2}$ so that its end shall fall upon the path L I L′ of the centre of the shadow at some point *g*, this line when produced backwards will meet the moon's disc at a point *d*, at which total darkness begins.

A similar line equal to $\frac{D-d}{2}$ drawn from M to fall upon the path of the shadow's centre at some point *ε* towards the end of the eclipse when produced backwards will find on the moon's disc some point *f* at which total darkness ends.

The method of projecting a lunar eclipse is adopted, with some variations, in the projections of a solar eclipse; the computations being for the apparent places of the sun and the moon, with the parallax applied to them, at times near the conjunction, on the day when a solar eclipse is expected to take place.

In this case the arc L I L′ in the figure, with necessary changes in position, etc., would represent the relative path of the moon's centre, the sun's disc being then considered to be fixed, at the centre of projection, with its radius, that of the third circle of rule (3). The radius of the second circle would also be changed, the moon's disc taking the place of the earth's shadow, as the coverer, the radius of the second circle of the projection would be half the sum of the diameter of the discs of the sun and the moon.

CHAPTER VII.

ON CONJUNCTIONS OF THE PLANETS CALLED GRAHA-YUTI.

Chapter VII. deals with conjunctions of the planets, which are called their fight, or association with each other, according to the degree of light which they emit.

Rule (2) refers to cases in which the times of conjunction may be past or future, in which the planet having the greater velocity may be in advance or behind the other; when the planets are moving eastward with a direct motion, or when one or both are moving with a retrograde motion.

Rules (3 and 4). A time is assumed, sufficiently near the conjunction, which each of two planets, for short intervals may be considered to be moving uniformly.

Let l_1 and l_2 be the longitudes of two planets A and B, whose latitudes are nearly the same at a given time, and whose daily motions at that time are m_1 and m_2 respectively, of which m_1 is greater than m_2; in the case in which the motions of both are direct, C B A. . . . . and the interval required is d in days, or fractions of a day, the required longitude of conjunction being l.

Rule (5). Then

$$\frac{m_1 (l_2 - l_1)}{m_1 - m_2} \text{ and } \frac{m_2 (l_2 - l_1)}{m_1 - m_2}$$

are called the changes of the planets which are to be added to the given longitudes if the conjunction is future, in which case

Rule (6) $$l = l_1 + \frac{m_1 (l_2 - l_1)}{m_1 - m_2} \text{ or } l_2 + \frac{m_2 (l - l_1)}{m_1 - m_2} = \frac{m_1 l_2 - m_2 l_1}{m_1 - m_2}$$

and the interval between the given time and the time of conjunction

$$d = \frac{l_2 - l_1}{m_1 - m_2}$$

Rule (7). Next, when the difference in latitude between the two planets is too great to be neglected.

The lengths of day and night of the places of the planets are to be found at the time of conjunction, their latitudes also in minutes, and their times from noon, and that for the rising and setting of each planet with the horoscope are to be computed.

A correction, called Drikkarma, is also requisite to be applied to the longitude of a planet for finding the point of the ecliptic (the Udaya Lagna) which rises simultaneously with a planet.

This correction consists of two parts, one called the Ayana and the other the Aksha Drikkarma.

These parts are differently estimated in different books. In the Surya Siddhanta to find the Aksha-Drikkarma we are told to—

Rule (8). "Multiply the latitude of the planet by the equinoctial shadow and divide the product by 12; the quantity obtained being multiplied by the time in Ghatikas from noon of the planet's *place* and divided by half the length of the day of the planets *place* gives the correction called the Aksha."

Rule (9). This correction is to be subtracted from the planet's place when east of the meridian, and the latitude of the planet is north; but it is to be added to the place when the latitude is south.

To find the correction called Ayana we are told to—

Rule (10). "Add 3 signs to the planet's place and find the declination from the sum. Then the number of minutes contained in the planet's latitude multiplied by the number of degrees contained in the declination gives in seconds the correction (called the Ayana-Drikkarma)."

Rule (11). The Ayana correction is to be added to or subtracted from the planet's place, according as the declination and the planet's latitude are of the same or different names.

The rules by which these two corrections were made would seem to have undergone considerable change from the original form in which they were constructed, Bhaskara, in the Siddhanta Siromani following Brahmegupta, gives rules for finding the difference in the times of rising of a planet and of the corresponding point of the ecliptic which determines the longitude of the planet.

This difference, as an entire correction in time, is found from two horary angles to which the names Ayana and Aksha Drikkarma corrections are given. If these angles be denoted by θ and ϕ respectively; and the latitude of the planet by λ, the latitude called the Spashta Sara (the rectified latitude) by λ^1, and the latitude of the observer's place be l, d being put for declination of the planet; then Bhaskara's rules give for the computation of the Drikkarma correction

$$\text{Sin } \theta = \frac{\sin \lambda}{\cos d} \times \text{sine Ayana Valana}$$

$$\text{Sin } \phi = \frac{R \sin \lambda^1}{\cos d \cos l} \times \text{sine Aksha Valana.}$$

In which are to be substituted the sines of the two Valanas which have been already given in the description of them in Chapter IV.

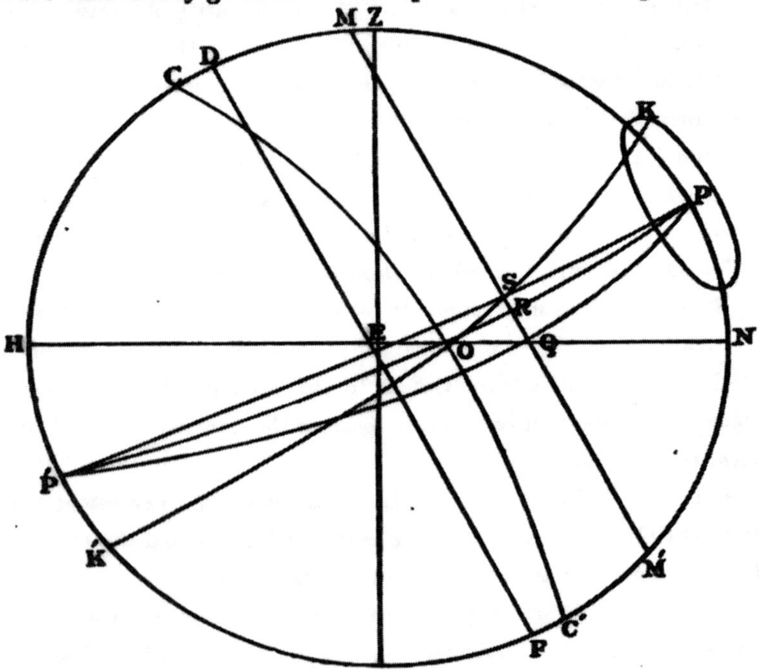

EXPLANATION.

The nature of these corrections may be explained as follows—

Let the above figure represent a projection of circles of the Eastern Hemisphere on the Meridian of the place, Z the zenith, H E N the horizon, P^1 its north point, D E F the equinoctial, E its east point, and P its pole, C O C^1 the ecliptic, and K its pole. S a planet or star, M S Q M^1 the diurnal circle through S. K S O K^1 the circle of latitude passing through S and meeting the ecliptic in the point O.

Therefore, O will be the point of the ecliptic, which determines the longitude of the planet at S, and O S will be its latitude (λ).

As represented in the figure, O is supposed to be in the horizon at a time after the rising of the planet through the arc Q S of the diurnal circle.

At this moment if great circles be supposed to be drawn from the pole P of the equinoctial, to pass through the three points S, O and Q, then the angle P O N becomes the Aksha Valana, and K O P becomes the Ayana Valana, and the sum or difference of these angles according to the position of the planet S (here taken as a sum), is the true Valana.

Now the time taken by the planet from the point Q to S of its diurnal circle is the horary angle Q P S expressed in time.

This angle consists of the two parts, S P O and O P Q, which have been here denoted by θ and ϕ respectively. And in the spherical triangle S P O we have $\frac{\sin S P O}{\sin S O P} = \frac{\sin S O}{\sin S P}$.

But O S is the latitude of the planet $= \lambda$, S P is its co-dec. $= 90 - d$ and the angle S O P is the Ayana Valana,

$$\therefore \quad \sin \theta = \frac{\sin \lambda}{\cos d} \times \sin \text{Ayana Valana.} \qquad (1)$$

Also in the triangle S O Q we, have, approximately

$\frac{\sin R Q}{\sin O R} = \frac{\sin R O Q}{\sin R Q O}$, in which O R is called the rectified latitude λ^1.

O Q R is nearly equal to the angle H E D or $= 90° - l$ and R O Q is the Aksha Valana, hence

$$\text{Sin Q R} = \frac{\sin \lambda'}{\cos l} \times \text{sin Aksha Valana} \qquad (2)$$

The arc of the equinoctial which corresponds to the arc Q R of the diurnal circle, and which measures the angle O P Q or ϕ, is found from

$$\text{Sin arc } \phi = \frac{\text{R sin Q R}}{\cos d} \qquad (3)$$

Substituting sin Q R from (2) in (3) we have

$$\text{Sin } \phi = \frac{\text{R} \sin \lambda'}{\cos l \cos d} \times \text{sin Aksha Valana} \qquad (4)$$

The angles or arcs θ and ϕ computed from (1) and (4), and expressed in time, by Asus reckoned each at one sixth of a sidereal minute, are the same as those given by the rules of Bhascara for finding the difference in time between the rising of a planet and the rising of the corresponding point of the ecliptic.

Rule (11) The Drikkarma correction is applied to the time of conjunction also of a planet with a star, of which the difference in latitude is too great to be neglected, also when finding the phases of the moon.

(12) It is likewise applied to the case of two planets, whose common longitude and apparent time of conjunction are determined by rule (6) of this chapter.

Rule (13) states the apparent diameters of the five planets Mars, Saturn, Mercury, Jupiter and Venus, to be respectively in Yojanas 30, 37½, 45, 52½ and 60.

(14) These magnitudes when reduced by rule 14, give their apparent diameters in arc of a great circle, 2′, 2½′, 3′, 3½′ and 4′.

Rule (15) Gives directions by which an observation may be made on a bright planet, or star, as shown by its reflection in a mirror.

We are told to fix a gnomon on a levelled floor, and to mark the shadow which it casts on the floor, a mirror is to be placed at the marked extremity of the shadow: "Then the planet will be seen (in the mirror) in the direction passing through the end of the shadow and the reflected end of the gnomon."

Rules (16), (17) give an imperfect description of the method of observing the two planets as seen at a conjunction.

Two styles are to be erected in a line in the north and south direction, each of five cubits in length with a cubit buried in the ground " at a distance equal to that of the two planets reduced to digits; the shadows are to be drawn from the bottoms of the styles, and lines drawn from the ends of the shadows to those of the styles: then the astronomer may show the planets in the lines, thus the planets will be seen in the heavens at the end of the styles."

The remaining verses from the 18th have reference to the various names given to the associations and fights of the planets, the kinds of fights, distinguishing which is the conqueror and which is conquered, etc., and in the last verse it is remarked that the associations and fights of the planets "are only imaginary, intended to foretell the good and evil fortune of people, since the planets being distant from each other move in their own separate orbit."

CHAPTER VIII.

ON THE CONJUNCTION OF PLANETS WITH STARS.

The chief object in Chapter VIII. is to find the apparent longitudes and latitudes of the principal stars of the 27 Asterisms or Nacshatras, into which the Hindu Ecliptic is divided, near which the planets may pass in their course through their respective orbits.

The apparent longitude of the principal star, or Yogatara, of an Asterism is not determined at once by the signs, degrees, minutes, etc., reckoned from the origin of the Ecliptic, but by the number of minutes of arc between the beginning, or first point of the Asterism, and the point of intersection of the Ecliptic, with a declination circle passing through the star; this arc is called the Bhoga of the Asterism, and the apparent longitude of the principal star is then found by adding the number of minutes in this arc to the longitude of the beginning of the Asterism.

The Bhoga is, therefore, only an apparent difference of longitude.

The Bhogas of all the 27 Asterisms are given with some differences in different Siddhantas; they are expressed by the number of minutes contained in them.

The apparent latitude of a star in Hindu astronomy is the arc of a *declination* circle measured from the star to the point of intersection of this circle with the Ecliptic.

As the apparent longitudes and latitudes of all the principal stars of the 27 Asterisms have been already fully given in the first part, as also of four other stars mentioned in this chapter, it is unnecessary to repeat them here.

CHAPTER IX.

ON THE HELIACAL RISING AND SETTING OF THE PLANETS AND STARS.

The Chapter begins by distinguishing between the rising and setting of Mercury and Venus (which are never very distant from the sun), from the rising and setting of the three planets, Mars, Jupiter and Saturn, whose longitudes may differ from that of the sun by as much as a semi-circle.

Rule (4). To find the time at which a planet rises or sets heliacally a day near the required time is chosen, and the true longitudes of the sun and the planet are to be found for this day.

The Drikkarma correction, as mentioned in Chapter VII., is then to be computed and applied to the place of the planet.

It has been before remarked that the difference in time between the rising of a planet and the rising of the corresponding point of the Ecliptic, which determines the longitude of the planet, is called the Drikkurma when expressed in time.

(5) The time in pranas between the rising of the point of the Ecliptic corresponding to the planet's place and the place of the sun is then to be found by rule 49, Chapter III.

This time in pranas, divided by 60, gives what is called the Kalansas, or time turned into degrees, at which, before sunrise, a body rises heliacally.

(6) The Kalansas for Mars, Jupiter, and Saturn are stated to be 11, 15 and 17 degrees of time respectively.

(7-8) When the motion of Venus or Mercury is retrograde, Venus is stated to rise or set heliacally by 8 degrees (of time) and Mercury by 12 degrees. But when the motion is direct, Venus rises or sets heliacally with 10 degrees (of time) and Mercury by 14.

(9) When the Kalansas of a planet found by the rule 5 are

greater than the numerical Kalansas mentioned above, the planet becomes visible, but it is invisible when the computed Kalansas are less.

(10) "Find the difference, in minutes, between the Kalansas (*i.e.*, Kalansas found from the place of the planet at the given time, and those which are the planet's own above mentioned); and divide it by the difference of the daily motions of the sun and the planet; the quantity obtained is the interval in days (Ghatikas, etc.) between the given time and that of the heliacal rising or setting. This holds when the planet is direct, but when it is retrograde, take the sum of the daily motions of the sun and the planet for the difference of the diurnal motions.

(11) "The daily motions of the sun and the planet, multiplied by the number of pranas contained in the rising periods of the signs occupied by the sun and the planet, and divided by 1,800, become the motions in time.

"From these motions (turned into time) find the past or future days, ghaticas, etc., from the given time to the time of the heliacal rising or setting of the planet."

In verses (12) to (15) the Numerical Kalansas of the principal stars of the 27 Asterisms are specified.

By rule (17) the Drikkarma is applied to their longitudes and through them the days past or future from the given time to the time of heliacal rising is found by means of the daily motion of the sun.

(18) Gives the names of a few stars which never set heliacally, as α Lyra, Capella, Arcturus, α Aquilæ, α Andromedæ, α Delphini.

CHAPTER X.

ON THE PHASES OF THE MOON AND THE POSITION OF THE MOON'S CUSPS.

The moon when rising or setting heliacally becomes visible in the western horizon according to the rules before mentioned; she is stated to become visible by 12° of time, and by the same number of degrees she becomes invisible in the eastern horizon.

On a day when the moon does not rise or set heliacally, the rule for setting on a given day in the light half of the Lunar month is—

(2) " Find (for sunset of that day) the true places of the sun and moon, and apply the two portions of the Drikkarma to the moon's place.

" From those places 'with 180° added' find the time in pranas (as directed in rule 5, Chapter IX.). At these pranas after sunset the moon will set."

The daily rising of the moon after the full requires a different rule.

(3) " Find the true places of the sun and moon at sunset and add 180° to the sun's place (and apply the two portions of the Drikkarma to the moon's place); from these places (*i.e.*, the sun's place with six signs added and the moon's place with the Drikkarma applied), find the time in pranas (as before directed rule (5), Chapter IX.) At this time in pranas after sunset the moon will rise."

The next four rules of the Chapter have reference to calculations necessary for the purpose of laying down lines, &c., which are used in the projection of the moon's phase on the given day, as set forth in rule (8).

(4) " Find the difference of the sine of declinations of the sun and the moon when they are of the same name (*i.e.* on the same side of the equinoctial), otherwise find the sum, to this result give the

name of the same direction south or north at which the moon is from the sun.

(5) "Multiply the result by the hypotenuse of the gnomonic shadow of the moon (as found in Chapter III.), find the difference between the product and twelve times the equinoctial shadow if the result be north, if it be south find the sum of them."

(6) The amount thus found divided by the sine of co-latitude of the place gives the Bahu or the base (of a right angled triangle); this is of the same name of which the amount is.

And the sine of the moon's altitude is the Koti (or perpendicular of the triangle). The square root of the sum of the squares of the Bahu and Koti is the hypotenuse (of the triangle).

To give these rules a modern form let D and d respectively be the declinations of the sun and moon; s the shadow of a style of 12 digits cast by rays of the moon; h the corresponding hypotenuse of the moon's shadow $= \sqrt{s^2+12^2}$; α the moon's altitude at the time.

$$\begin{aligned} \text{Then Bahu} &= \frac{(\sin D \pm \sin d)\, h \mp 12\, p}{\text{cosin latitude}} \\ \text{Koti} &= \sin \alpha \qquad\qquad \text{(A)} \\ \text{Hypotenuse} &= \sqrt{\text{Bahu}^2 + \text{Koti}^2} \end{aligned}$$

(7) "Subtract the sun's place from that of the moon. The minutes in the remainder divided by 900 give the illuminated part of the moon. This part multiplied by the moon's disc (in minutes) and divided by 12 becomes the Sphuta, or rectified illuminated part."

If L and l be taken to denote the longitudes respectively of the sun and the moon

$$\text{The Sphuta} = \frac{l' - L'}{10{,}800'} \times \text{moon's disc} = \frac{l^\circ - L^\circ}{180^\circ} \times \text{moon's disc} \quad \text{(B)}$$

(8) To project the phase of the moon "(on a board or levelled floor), having marked a point representing the sun, draw from that point a line equal to the Bahu, in the same direction in which the Bahu is, and from the end of the Bahu a line (perpendicular to it equal to the Koti to the west, and draw the hypotenuse between the end of the Koti and the point (denoting the sun).

(9) "About the point where the Koti and the hypotenuse meet describe the disc of the moon."

In this disc suppose the directions (east and west) through the line of the hypotenuse.

To represent the projection thus described, assume a point S and a horizontal line S N, in which S N = computed value of Bahu (A), and N M perpendicular to N S = Koti, sine of moon's altitude join M S. At M a circle is described representing the moon's disc meeting the hypotenuse M S in *e* and *w*, the east and west points, a line at right angles to M S through M will cut the disc in the points *n* and *s*, the north and south points.

The solar rays in the direction S M will illuminate the hemisphere M *n e s*, which is turned towards the sun, and if *e o* be the part of

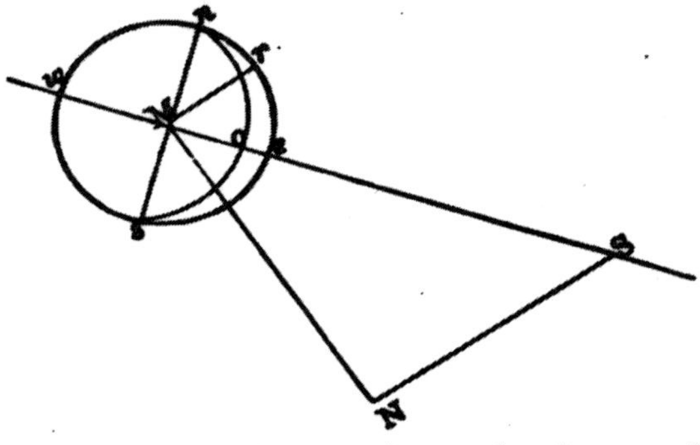

the hypotenuse called the Sphuta, and *n o s* be the arc of a circle passing through the three points *n*, *o* and *s*, then the crescent or line *n o s e* will be the illuminated part seen from the earth.

The direction of the horns of the crescent is marked by the line *n s* through their extremities, and the inclination of this line to the horizontal direction is the angle *n* M *r*; it is equal to the angle S M N, which the hypotenuse S M makes with the perpendicular N M.

It may be observed that in the projection of solar and lunar eclipses, and of the phases of the moon when straight lines are referred to, such as sines, cosines, diameters of discs of the sun and the moon, etc., they are estimated in minutes of arc of any circle, which is assumed as the foundation of the scale of projection; the circumference always consisting of 21,600 minutes of arc, of which the radius is 3,438 minutes, and the digit is 70 minutes.

CHAPTER XI.

ASTROLOGICAL INTERPRETATIONS.

Treats of rules for finding the times at which the declinations of the sun and moon are equal, and the purposes of the Chapter are purely astrological in character.

A fire called Pata is supposed to be produced by a mixture of the solar and lunar rays in equal quantities, and burnt by the air called Pravahu.

The Pata is personified as a horrible monster, black in colour, hard bodied, red eyed, and gorbellied, of a malignant nature, producing evil to mankind, and destroying the people.

It occurs frequently when the declinations of the sun and moon become equal.

First: When both bodies are on the same side of the equator, and the sum of their longitudes is equal to 12 signs or 360°.

Secondly: When they are on opposite sides of the equator and the sum of their longitudes is 6 signs or 180°.

Rules are given for the times when these occurrences take place, indicating when they are in the past or in the future; whether they happen before or after midnight; their duration from the beginning to the end is a horrible interval, during the continuance of which all rites are prohibited; and it is of advantage to know these times, for virtuous acts, for purposes of bathing, almsgiving, prayers, funeral ceremonies, religious obligations, burnt offerings, etc.

In addition to the above, there are other frightful periods, when all joyful acts are prohibited; the Vyatipas, Bhasandhis and Gandantas.

If, when the minutes contained in the sum of the longitudes of the sun and the moon are divided by 800 (*i.e.*, the minutes in 13° 20′, the

V

extent of a Nacshatra), the quotient should be between the numbers 16 and 17, the Pata occurs called Vyati-Pata.

Again, the last quarters of the three Nacshatras, Aslesha, Jyestha and Aswini, are called the Bhasandhis.

And the first quarters of each of the three Nacshatras following, namely, Magha, Mula and Aswini, are called the Gandantas.

CHAPTER XII.

ON COSMOGRAPHICAL THEORIES OF THE HINDUS.

From verses (1) to (9) a series of questions are proposed about the earth, its magnitude, its form and divisions. The situation of the seven Patala Bhumis or imaginary lower regions of the earth.

Questions also regarding the sun's revolutions, the causes of day and night of the Gods, the Demons and the Pitris. On the order of the stars, and planets, the position of their orbits with respect to each other in the Universe, etc., which are answered in subsequent verses.

The verses from (10) to (32) relate to imperceptible agencies of creation, but it is not in the plan of this work to describe the Metaphysical theories of the Hindus, regarding the creation of the Universe which may be found in the Vedas the Puranas and other works. Nor will it be necessary to dwell on the peculiar Geographical theories detailed in verses from (33) to (54), some of which are purely figments of the imagination, and of the remainder the more important parts have been already sufficiently discussed in foregoing Chapters.

Verses from (55) to (74) have reference principally to day and night at different places on the earth easily deducible from a knowledge of the circles of the sphere, and the apparent motions of the sun and moon. Such as the day and night at places on the equator, at the tropics, and at the poles.

The increase or decrease to day or night caused by the varying positions of the sun in an oblique sphere at places within the tropics.

Places on the earth at which some signs are always visible and others always invisible.

At the poles the sun is above the horizon for half the year, and invisible for the remaining half, and at a pole the direction of the Gnomonic shadow points always from the pole.

v 2

To a person proceeding northward the altitude of the pole increases with the latitude of the place.

The starry sphere is said to revolve constantly through the influence of the Pravaha winds, as also do the planets confined within their respective orbits.

The Pitris, situated in the upper part of the moon, behold the sun throughout a fortnight.

ON THE BRAHM-ANDA.

The Brahm-Anda or the golden egg of Brahma is the vast hollow sphere of the universe at the centre of which is the earth. And within it all the stars are supposed to revolve daily; beneath them are the orbits of the planets Saturn, Jupiter, Mars, the Sun, Venus, Mercury, and the Moon, in the order of their distances from the centre.

The orbit of the Moon which is the smallest, is estimated to have a circumference equal to 324,000 yojanas, which as a mean is a fairly

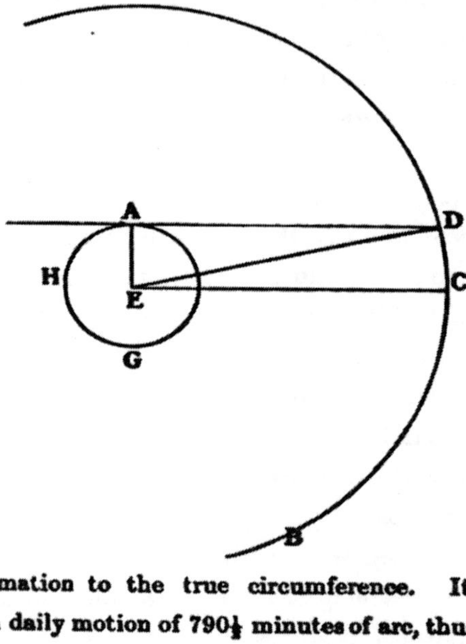

good approximation to the true circumference. It was deduced from its mean daily motion of 790½ minutes of arc, thus, in the figure

Let E be the centre of the earth A H G, and B C D a portion of the moon's orbit.

Also let A D and E C represent the sensible and rational horizons of an observer at A.

Then C D will be approximately equal to E A, the radius of the earth, which is reckoned in the Surya Siddhanta to be 800 yojanas.

Again, the moon's horizontal parallax A D E, or D E C, is estimated by the Hindus to be 52′ 42″ of arc, and

$$\frac{\text{The moon's daily motion in Yojanas}}{\text{C D in Yojanas}} = \frac{\text{Daily motion in minutes}}{\text{C D in minutes}}, \text{ or}$$

$$\frac{\text{Daily motion in Yojanas}}{800\text{ Y}} = \frac{790{\cdot}5'}{52{\cdot}7'} = 15,$$

Therefore, the moon's daily motion in her orbit $= 12{,}000$ Yojanas.

But reckoning the sidereal period of the moon, or the time of her revolution round the earth to be 27 days, the circumference of the circle of the moon's orbit would be $27 \times 12{,}000 = 324{,}000$ Yojanas.

And this formed the foundation for finding the circumferences of the other planets by the false hypothesis, which was accepted by all astronomers before the time of Kepler, by which it was assumed that the planets moved each in its own orbit with the same velocity of all the others, the differences in their sidereal periods being accounted for by the greater circumferences to be traversed by the more distant planets than by those which were nearer.

In the Hindu astronomy, as in the case of the moon, every planet was supposed to traverse nearly 12,000 Yojanas of its orbit daily, one-fifteenth of this being the semi-diameter of the earth, and one-fifteenth of the daily motion in arc being the planet's horizontal parallax.

The circumference of a circle, called the middle circle of the starry sphere, which is supposed to revolve about all the planets, is found by multiplying the sun's orbit in Yojanas by 60

$$= 4{,}331{,}500 \times 60$$
$$= 259{,}890{,}000.$$

The circumference of the sphere of the Brahm-Anda to which the solar rays extend, is declared to be equal to the product of the moon's revolutions in a Kalpa (57,753,336,000) by the circumference of the moon's orbit (324,000).

The dimensions of the orbits of the planets, etc., have been arranged as given in this chapter in the following order—

The Moon	324,000	Yojanas.
The Sighrochcha (apogee) of Mercury	1,043,209	„
Sighrochcha „ of Venus ..	2,664,637	„
Sun, Mercury and Venus	4,331,500	„
Mars	8,146,909	„
Jupiter	51,375,764	„
Saturn	127,668,255	„
Sphere of the stars (circumference)	259,890,012	„
Sphere of the Brah Mandee (circ.) ..	18,712,080,864,000,000	„
The Moon's Apogee	38,328,484	„
The Moon's ascending Node ..	80,572,864	„

CHAPTER XIII.

ON THE CONSTRUCTION OF THE ARMILLARY SPHERE AND OTHER INSTRUMENTS.

At the beginning of this Chapter from verses (3) to (12), directions are given for the construction of a Golayantra or Armillary Sphere.

A wooden terrestrial globe is prepared for its centre, having an axis projecting to two supporting circles, representing the equinoctial and solstitial colures.

To the supporting circles is fixed a circle representing the equinoctial, and parallel to it small circles are arranged through the ends of Aries, Taurus and Gemini in the northern hemisphere and through the ends of Libra, etc., of the southern, serving as diurnal circles of the 12 signs.

Similar small diurnal circles are fixed to the supporting circles for some of the principal stars, as for Abhijit (α Lyra), the Rishis or seven saints (stars of Ursa Major), Agastya (Canopus), Brahma (Auriga), etc.

The position of the two solstices are to be marked on one of the supporting circles at the distance of the sun's greatest northern and southern declinations.

On the other supporting circle the positions of the two equinoxes are to be marked at the intersecton of it with the equinoctial.

Strings are to be stretched joining the equinox with each of the signs at every arc of 30°, as if it were intended to show the plane of the ecliptic.

And the ecliptic itself is to be formed by a circle passing from solstice to solstice.

These hints, are nearly all that can be gathered regarding the construction of the armillary sphere, as described in the Surya Ṣiddhanta.

In the Surya Siddhanta several instruments are mentioned for measuring time.

A self-revolving sphere is to be made with its axis directed to the poles. The lower part of it is to be covered by wax cloth, and it is to be made to rotate by the force of a current of water for the knowledge of the passage of time.

Other self-acting instruments are to be made, but the method of construction is to be kept secret. The application of some of the methods is said to be difficult of attainment.

A wheel with hollow spokes half filled with mercury, or water, or a mixture of oil and water.

The hour is to be known also from the instruction of the teacher by means of the Gnomon, the staff and circle in various ways, and by mercury and sand.

Or the hour is to be known by the Kapala Yantra or Clepsydra. It is "a copper vessel, shaped as the lower half of a water jar; it has a hole in its bottom, and being placed upon clean water in a basin sinks exactly 60 times in a Nycthemeron."

CHAPTER XIV.

VARIOUS KINDS OF TIME.

In this Chapter are described the nine kinds of time called Manas, which are named the Brahma, the Divya, the Pitrya, the Prajapati, and those that relate to this world, the Surya or Solar, the Lunar, the Sidereal, the Terrestrial, and that of Jupiter for knowing the Samvatsaras.

The Solar Mana is that by which are determined the lengths of day and night, the Shadasiti-Mukhas, the solstitial and equinoctial times and the holy days of Sankranti on which good actions bring good desert to the performer.

Verses from (4) to (6) relate to a peculiar division of time consisting of successive periods of 86 solar days beginning from the time when the sun enters the sign of Libra; the 86th day of each period is called Shadasiti-Mukha, and there are four such days in the year, the first happens when the sun is at 26° of Sagittarius, the second when he reaches 22° Pisces, the third when he is at 18° Gemini, and the fourth at 14° Virgo. The remaining 16 Saura days or degrees of the Saura month, when the sun is in Virgo are sacred, good actions performed in those days confer great merit, equal to that of a sacrifice, a gift then in honour of deceased ancestors is imperishable.

Verse (7) refers to the equinoxes as being diametrically opposite in the middle of the starry sphere; so are the two solstices.

(8.) The beginnings of the four signs Taurus, Leo, Scorpio and Aquarius, are called Vishnu-padi or feet of Vishnu.

The sun's progress northward from his entrance into Capricorn through six signs is called the Uttar-Ayana or northing, and from the entrance into Cancer the progress is called Dakshin-Ayana, or the southing of the sun.

10. From the winter solstice, the periods during which the sun remains in two signs are the seasons named successively--

1.	Sisira (very cold).	2.	Vasanta (spring).
3.	Grishma (hot).	4.	Varsha (rainy).
5.	Sarat (autumn).	6.	Hemanta (cold).

The holy time called Sankranti or the time of the sun's entering a sign, determines the (Saura) Solar or Sidereal month, during which the sun passes through each arc of 30° from the beginning of one sign to that of the next.

Therefore, Saura or solar months, each consisting of 30 Saura days or degrees, are of unequal length reckoned in mean solar days on account of the unequal motion of the sun in the ecliptic, but the aggregate is equal to the sidereal year, which in the Surya Siddhanta is reckoned to be 365 days 6 hours 12 minutes 36·56 seconds, and the mean Saura, or solar month, would therefore be 30 days 10 hours 31 minutes 3·5 seconds.

The Saura month of greatest length is Ashadha, consisting of 31 days 14 hours 39 minutes 7 seconds; and the least is Pausha, which is 29 days 8 hours 21 minutes 7 seconds.

The lunar months are named from the Nacshatras, in which the moon happens to be on the 15th day of such months.

The first lunar month is Chaitra from the Nacshatra Chitra; the second is Vaisakha from Visakha; the third, Jyeshtha, from Jyeshtha; the fourth, Ashadha, from Purvashadha; the fifth, Sravana, from Sravana; the sixth, Bhadrapada, from Purva Bhadrapada; the seventh, Aswina, from Aswini; the eighth, Kartika, from Krittika; the ninth, Margasirsha, from Mrigasirsha; the tenth, Pausha, from Pushya; the eleventh, Magha, from Magha; and the twelfth, Phalguna, from Purva-Phalguni.

As time in the abstract is in duration the same for all measures of it, the term Mana, or kind of time, can only have reference to the origin from which each specific unit is derived.

The only invariable astronomical unit, as far as we know, is the sidereal day, or the time of one rotation of the earth about its axis, or the time of one apparent revolution of the sphere of the stars about the earth.

The solar day from sunrise to sunrise, the lunar day, the solar month in its different forms, the lunar month from one full moon to another, etc., are all variable magnitudes, which in their mean values are referred for comparison to the invariable sidereal time.

The Manas named at the beginning of this Chapter have in other ways been mentioned in former parts of the work. Here it would appear they have more particular reference to their uses in religious observances, holy times of sacrifice, etc.

The Mana of Brahma is the Kalpa.

The Mana of Prajapati (the father of Manu) is the duration of 71 Maha Yugas.

The Mana of the Gods is their day and night, or a year of mortals.

The Mana of the Pitris is the lunar month, the duration of their day and night.

The lunar month is again the lunar Mana.

The sidereal Mana is the sidereal day.

The years of Jupiter are named by analogy from lunar months when Jupiter rises or sets heliacally.

CHAPTER XV.

CONCLUDING OBSERVATIONS.

The purpose of the writer has now been accomplished, namely, to place before the reader some simple account of the nature and peculiarities of Hindu Astronomy.

No doubt much has been omitted which might have been advantageously inserted for a complete appreciation of the subject, but it is hoped that sufficient has been stated to present a general sketch which may enable those interested to retain a grasp of its principal features. It may, however, be desirable before leaving the subject, to offer a few remarks even at the risk of repetition.

The author's object has been, in the first place, to point in some measure to, and emphasize, the extreme antiquity of the science of Astronomy, as found in India: secondly, to give such a description as to enable the general reader to note not only the similarities to, but also the differences from, the astronomical science of the West, with a view, by such comparison, to form his own estimate of the origin of the one system, or of the other: thirdly, to show that even in the Paganism and mythology of the Hindus there is a substratum of worth so far as these are connected with their system of Astronomy.

Upon the first point (the antiquity of that system), it may be remarked, that no one can carefully study the information collected by various investigators and translators of Hindu works relating to Astronomy, without coming to the conclusion that, long before the period when Grecian learning founded the basis of knowledge and civilization in the West, India had its own store of erudition. Master minds, in those primitive ages, thought out the problems presented by the ever recurring phenomena of the heavens, and gave birth to the ideas which were afterwards formed into a settled system for the use and benefit of succeeding Astronomers, Mathematicians, and Scholiasts, as well as for the guidance of votaries of religion.

No system, no theory, no formula, concerning those phenomena could possibly have sprung suddenly into existence at the call, or upon the dictation of a single genius. Far rather is it to be supposed that little by little, and after many arduous labours of numerous minds, and many consequent periods passed in the investigation of isolated phenomena, a system could be expected to be formed into a general science concerning them.

Further, as Bailly has truly remarked, Astronomy cannot be numbered among those arts and sciences which in a more peculiar manner belong to the sphere of imagination, and which by the wonderful energy of vigorous and splendid genius are often brought rapidly to perfection. It is, on the contrary, by very slow advances that a science founded upon the basis of continued *observations*, and profound mathematical researches, approaches to any degree of maturity. Many ages, therefore, must have elapsed before the motion of the sun, moon and planets could be ascertained with exactness; before instruments were invented to take the height of the pole and elevation of the stars; and before the several positions in the heavens could be accurately noted on descriptive tables or a celestial globe.

It is in the light of such considerations as these, that the investigator of the facts relating to Hindu Astronomy is compelled to admit the extreme antiquity of the science. As far back as any historical data, or even any astronomical deduction, can carry the mind, the conception of the ecliptic and the zodiac is presented to view in a system. It is very reasonable therefore to infer that an unknown number of centuries must have elapsed previously, during which the primitive philosophers established their ideas in the connected manner indicated in the conception referrèd to. The same observation and inference may be applied to the reasoning powers brought into play in the science of mathematics and kindred subjects, in which even in their most abstruse aspects, the Hindus, at any rate amongst the higher and more educated castes, have shown a deeply reflective capacity. In some quarters, an attempt has been made to minimise these faculties upon grounds which, in the opinion of the present writer, are not only inadequate, but which show in the critics themselves a want of appreciation of the true merits of Hindu Astronomy. An impartial investigation of the crcumstances relating to the question whether the Grecian Astronomy (which is the parent of our system), was original in its nature, and was copied by the Hindus, places it beyond doubt that the Hindu system was essentially different from and independent of the Greek.

Some of the dissimilarities, as well as some of the similarities in the two systems have been shown in the preceding chapters. As to the former, it may be truthfully asserted that nothing like the *fixed* ecliptic with its fixed concomitant arrangement of lunar Asterisms is to be found in the Ptolemaic and later systems. Neither do we find in these latter, anything like the method employed by the Hindus in estimating long periods of time, nor that of determining longitudes of the sun, moon, and planets from their position in a Nacshatra. Moreover, it is only necessary to refer to the method adopted by Hindu Astronomers for determination of longitudes by the calculated rising of the signs, and used also in finding the horoscope, and the nonagesimal point, and the culminating point on the ecliptic; there is no such method in our system. Even the process of calculations employed in regard to everything stated in Siddhantas, appears to exhibit a fundamental difference in the Hindu system, from processes employed in the science of the West. Again, it may be asked, where is there anything similar to the Palabha, or equinoctial shadow of the gnomon, used in that system, as an equivalent for the latitude of a place, and where is there anything like the formula entitled the Valana, in the projection of an eclipse?

Further, is there anything with us corresponding to the Hindu radius, estimated in 3,438 minutes of arc? These are unique, and go far to establish the contention that, whatever be the origin of the Hindu system, it certainly was not, in these and other particulars, copied from the Grecian or any European system.

Lastly, it has been the author's desire, by the preceding explanations, to dispel some of the supercilious ridicule cast by some Western critics upon Hindu methods of dealing with astronomical time, and upon their mythology. Such ridicule would appear to be unmerited, since the subject of it has been misunderstood. So far from the extraordinary numbers of years employed in computation by Hindu Astronomers being absurd, it has been shown that they were absolutely necessary to their peculiar system and methods, for ensuring accuracy. The astronomical mythology, likewise, of the Hindus, grotesque and barbarous as some of their stories may appear, had within it much that was valuable in point of instruction. No nation in existence can afford to compare its latter day tenets of science with its earliest theories and cosmography, without a smile at the expense of ancestors; but the Hindus, in this view, may, with not a little justifiable pride, point to their sciences of Astronomy, of Arithmetic, Algebra, Geometry and even of Trigonometry, as containing within them evidences of traditioned civilization comparing favourably with that of any other nation in the world.

APPENDIX I.

With regard to the supposed actual observations of the planets by the Hindu Astonomers at the epoch of the Kali Yuga, Laplace, after speaking of the Chinese and their scrupulous attachment to ancient customs which extended even to their astronomical rules, and has contributed among them to keep this science in a perpetual state of infancy, proceeds thus in his "Exposition du Systeme du Monde":—

"The Indian tables indicate a much more refined astronomy, but everything shows that it is not of an extremely remote antiquity. And here, with regret, I differ in opinion from a learned and illustrious astronomer (M. Bailly) who, after having honoured his career by labours useful both to science and humanity, fell a victim to the most sanguinary tyranny, opposing the calmness and dignity of virtue to the revilings of an infatuated people, who wantonly prolonged the last agonies of his existence.

"The Indian tables have two principal epochs, which go back, one to the year 3102 the other to the year 1491 before the Christian Era. These epochs are connected with the mean motions of the sun, moon, and planets, in such a manner that one is evidently fictitious; the celebrated astronomer above alluded to, endeavours in his Indian astronomy to prove that the first of these epochs is grounded on observation. Notwithstanding all the arguments brought forward with the interest he so well knew how to bestow on subjects the most difficult, I am still of opinion that this period was invented for the purpose of giving a common origin to all the motions of the heavenly bodies in the zodiac.

"In fact, computing, according to the Indian tables from the year 1491 to 3102, we find a general conjunction of the sun and all the planets, as these tables suppose, but their conjunction differs too much from the result of our best tables to have ever taken place, which shows that the epoch to which they refer was not established on observation.

"But it must be owned that some elements of the Hindu astronomy seem to indicate that they have been determined even before the first epoch. Thus the equation of the sun's centre, which they fix at 2·4173°, could not have been of that magnitude, but at the year 4300 before the Christian era.

"The whole of these tables, particularly the impossibility of the conjunction at the epoch they suppose, prove on the contrary that they have been constructed, or at least rectified, in modern times.

"Nevertheless, the ancient reputation of the Indians does not permit us to doubt that they have always cultivated astronomy, and the remarkable exactness of the mean motions which they assign to the sun and moon necessarily required very ancient observations."

It would appear from a paper on the "Trigonometry of the Brahmins," published in the "Transactions of the Royal Society of Edinburgh," Vol. IV. (1798), eight years after his first paper on "The Astronomy of the Brahmins," that Playfair was induced to modify his opinion with regard to Bailly's belief as to the origin of the Kali Yuga, to which he had referred in the construction of the Indian astronomical tables.

He says, "I cannot help observing, in justice to an author of whose talents and genius the world has been so unseasonably and so cruelly deprived, that his opinions, with respect to this era, appear to have been often misunderstood.

"It certainly was not his intention to assert that the Kali Yuga was a *real* era, considered with respect to the mythology of India, or even that at so remote a period the religion of Brahma had an existence.

"All I think Bailly meant to affirm, and certainly all that is necessary to his system, is that the Kali Yuga, or the year 3102 before our Era, marks a point in the duration of the world before which the foundations of astronomy were laid in the east, and those observations made from which the tables of the Brahmins have been composed."

APPENDIX II.

There are innumerable stars which, as far as we know, never change their relative situations, in consequence of which they are said to be fixed. Thus, three stars always form the same triangle, and with a fourth the same trapezium, and the manifold figures, which they may be conceived to represent when they are supposed to be joined by spherical arcs, have ever retained the same form and situation, or nearly so, since creation, and may continue so through endless time.

This fixity of character of the stars was recognised in the most remote ages, and with the Hindus it was the foundation upon which their system of astronomy was built. With them the path of the sun (the ecliptic) has its position also sensibly fixed with reference to the stars, although this is not the case with the great circle called the equinoctial, which has not that immoveable character. The celestial equator is continually changing in position, and the co-ordinates of the stars which are referred to it, that is, their Right Ascensions and Declinations, undergo changes yearly of a complex nature, whereas their changes in longitude are all of a character appreciably simple. The apparent slow motion of the equinoctial and solstitial points along the ecliptic, technically called the precession, is really a retrogression, by means of which all stars appear to move backwards at a mean annual rate of 50·1 seconds, causing an annual augmentation to their Longitudes of the same amount, so that if we have a table of Longitudes of stars for any one year (as for the beginning of the century 1800) then the mean Longitude for any other time may be found by simply adding or subtracting 50·1 seconds to each, for each succeeding or preceding year.

Again, the changes in the Latitudes of stars are so minute that some writers have supposed the Latitudes to be invariable; this, however, is not quite true, for from an examination of many of the principal stars, and by a comparison of their Latitudes after long intervals of time, it is found that some almost insensible changes do take place. Thus, out of a number of stars whose Latitudes were examined in 1815 and compared with those of the same stars as given for the year 1756, it was ascertained that in no case had the Latitudes altered annually by so much as ·32 of a second, and in some the changes were almost inappreciable. So that in a table of Latitudes and Longitudes, rectified for the beginning of a

W 2

century, the Latitudes may in general be depended upon within less than half a minute during the century, and the Longitudes at any time by applying the correction for precession.

The following table of zodiacal stars is taken from Dr. Gregory's Astronomy, 1802. By means of this table and a table of the moon's Longitude, he says, we may ascertain how often a given fixed star may be eclipsed by the moon in a given year. It will even be found useful now for projecting the position of the ecliptic on photographic charts of zodiacal stars at the present day. For if two prominent stars of the photograph be recognised whose latitudes are known, then circles may be conceived drawn about each star at distances equal to their Latitudes, estimated according to the scale of the photograph, and they will have a portion of the ecliptic as a common tangent; the points of the contact being points of the ecliptic having the same Longitudes as the stars. Thus, they serve for determining the Longitudes and Latitudes of the other stars of the photograph.

The ordinary phenomena of the solar system, such as eclipses of the sun and moon, the numerous occultations of planets and fixed stars, their conjunctions and oppositions, all occur either on the ecliptic or within a few degrees of it, and in a clear sky they may in general be observed with the unaided eye. To a diligent student of astronomy the order of their occurrence soon becomes familiar, and by aid of the table a simple calculation will give the time and position of each in succession.

***Table of the Longitudes, Latitudes and Magnitudes of the most remarkable fixed stars that the Moon can eclipse or make a near appulse to, exactly rectified to the beginning of the year* 1800.**

	Name of Star.	Mag.	Longitude.				Latitude.					Name of Star.	Mag.	Longitude.				Latitude.			
			s.	°	′	″	°	′	″					s.	°	′	″	°	′	″	
δ	Piscium	4	0	11	21	34	2	9	44	N	α	Libræ	2	7	12	17	52	0	21	48	N
ε	Piscium	4	0	14	44	33	1	5	37	N	12	Libræ	3	7	18	13	0	1	49	14	S
ζ	Piscium	4	0	17	4	48	0	13	11	S	γ	Libræ	3	7	22	20	32	4	24	41	N
δ	Arietis	4	1	18	3	7	1	48	7	N	η	Libræ	4	7	24	34	56	4	2	52	N
η	Tauri (*Alcyone*)	3	1	27	12	7	4	1	36	N	χ	Libræ	4	7	24	58	10	0	1	1	N
γ	Tauri	3	2	3	0	23	5	45	30	S	ϑ	Libræ	4	7	27	4	35	3	29	24	N
ε	Tauri	3	2	5	40	0	2	35	37	S	λ	Libræ	4	7	27	41	9	0	6	53	N
α	Taúri (*Aldebaran*)	1	2	6	59	43	5	29	2	S	δ	Scorpii	3	7	29	44	51	1	57	17	S
β	Tauri	2	2	19	47	0	5	21	59	N	π	Scorpii	3	8	0	8	58	5	26	15	S
ζ	Tauri	3	2	21	59	38	2	13	29	S	β	Scorpii	2	8	0	23	56	1	2	18	N
η	Geminorum	4	3	0	39	0	0	55	4	S	υ	Scorpii	4	8	1	51	13	1	39	52	N
μ	Geminorum	3	3	2	30	25	0	50	34	S	σ	Scorpii	4	8	5	0	36	4	0	23	S
γ	Geminorum	2	3	5	18	43	6	46	12	S	α	Scorpii (*Antares*) .. .	1	8	6	58	21	4	32	17	S
ε	Geminorum	3	3	7	8	53	2	2	28	N	τ	Scorpii	4	8	8	40	5	6	5	21	S
δ	Geminorum	3	3	15	43	52	0	12	19	S	γ	Sagittarii	3	8	28	28	26	6	56	48	S
β	Geminorum (*Pollux*) ..	1	3	20	27	57	6	40	4	N	μ	Sagittarii	4	9	0	25	21	2	22	24	N
γ	Cancri	4	4	4	15	14	3	10	22	N	λ	Sagittarii	4	9	3	31	54	2	5	31	S
δ	Cancri	4	4	5	55	32	0	4	13	N	φ	Sagittarii	3	9	7	33	11	3	55	22	S
ζ	Leonis	4	4	13	51	46	3	1	57	S	σ	Sagittarii	3	9	9	35	40	3	24	55	S
ο	Leonis	4	4	21	28	1	3	46	1	S	τ	Sagittarii	3	9	12	2	25	5	2	33	S
η	Leonis	4	4	25	6	46	4	51	9	N	ο	Sagittarii	3	9	12	11	59	0	53	36	N
α	Leonis (*Regulus*)	1	4	27	3	12	0	27	27	N	π	Sagittarii	4	9	13	27	44	1	28	7	N
ρ	Leonis	4	5	3	35	48	0	8	29	N	β	Capricorni	3	10	1	15	18	4	36	46	N
τ	Leonis	4	5	18	43	10	0	31	21	S	ε	Capricorni	4	10	17	24	23	4	57	31	S
υ	Leonis	4	5	22	14	55	3	2	51	S	γ	Capricorni	4	10	18	59	16	2	32	6	S
β	Virginis	3	5	24	19	10	0	41	36	N	δ	Capricorni	3	10	20	44	28	2	33	40	S
ο	Virginis	3	6	0	34	33	5	4	42	N	ι	Aquarii	4	10	25	55	40	2	3	47	S
η	Virginis	3	6	2	2	39	1	22	24	N	θ	Aquarii	4	11	0	28	5	2	43	22	N
γ	Virginis	3	6	7	23	4	2	48	57	N	λ	Aquarii	4	11	8	47	6	0	22	57	S
α	Virginis (*Spica*) ..	1	6	21	3	13	2	2	11	S	φ	Aquarii	4	11	14	20	58	1	3	8	S

8 888006 778916